ELEMENTS OF ELECTRONICS
BOOK 1

The Simple Electronic Circuit
and
Components

ALSO BY THE SAME AUTHOR

BP63 Alternating Current Theory (Elements of Electronics – Book 2) UR

BP64 Semiconductor Technology (Elements of Electronics – Book 3)

BP77 Microprocessing Systems and Circuits (Elements of Electronics – Book 4)

BP89 Communication (Elements of Electronics – Book 5)

BP111 Audio (Elements of Electronics – Book 6)

BP53 Practical Electronics Calculations and Formulae

BP144 Further Practical Electronics Calculations and Formulae

ELEMENTS OF ELECTRONICS
BOOK 1

The Simple Electronic Circuit
and
Components

by
F. A. WILSON
C.G.I.A., C.Eng., F.I.E.E., F.I.E.R.E., F.B.I.M.

BERNARD BABANI (publishing) LTD
THE GRAMPIANS
SHEPHERDS BUSH ROAD
LONDON W6 7NF
ENGLAND

PLEASE NOTE

Although every care has been taken with the production of this book to ensure that any projects, designs, modifications and/or programs etc. contained herein, operate in a correct and safe manner and also that any components specified are normally available in Great Britain, the Publishers do not accept responsibility in any way for the failure, including fault in design, of any project, design, modification or program to work correctly or to cause damage to any other equipment that it may be connected to or used in conjunction with, or in respect of any other damage or injury that may be so caused, nor do the Publishers accept responsibility in any way for the failure to obtain specified components.

Notice is also given that if equipment that is still under warranty is modified in any way or used or connected with home-built equipment then that warranty may be void.

First Published – September 1979
Reprinted – January 1984
Reprinted May 1987

British Library Cataloguing in Publication Data
Wilson, F A
Elements of electronics
Book 1
1. Electronic apparatus and appliances
I. Title
621.381 TK7870

ISBN 0 900162 82 1

Printed and bound in Great Britain by Cox & Wyman Ltd, Reading

The thirst to know and understand,
A large and liberal discontent;
These are the goods in life's rich hand,
The things that are more excellent.

Sir William Watson

CONTENTS

Page

INTRODUCTION

The aim of this book and of those which follow in the series can be stated quite simply – it is to provide an inexpensive introduction to modern electronics so that the reader will start on the right road by thoroughly understanding the fundamental principles involved. Thus those who ultimately make some branch of electronics their career will be that much more proficient, equally those who simply enjoy electronics as a hobby may derive more satisfaction especially from a greater appreciation of material in electronics periodicals. That the books are inexpensive is perhaps obvious. By accepting their small size the total information is that which in normal size textbooks would cost some 2–3 times as much.

Although written especially for readers with no more than ordinary arithmetical skills, the use of mathematics is not avoided as so often happens, for thereby one never really understands the subject completely. All mathematics required is taught as the reader progresses and it is hoped that those who perhaps have never had an interest in, nor even seen the need for the subject, may realise its usefulness and even find it stimulating. Apart from the fact that for some, a little expertise may make all the difference between finding a job or not, there is little doubt that some skill in electronics and mathematics will be an asset to almost everybody in the future, destined as we are to be controlled more and more by computers, microprocessors and the like as time progresses. There is also much to offer those who need revision or to be brought up to date with fundamentals, those with grown-up children asking awkward questions or preparing for entry to a college, or even those who may find a different approach refreshing.

The course is not written to a specific college or course syllabus, but concentrates on the understanding of the important concepts central to electronics rather than continually digressing over the whole field on the basis that once the fundamentals are mastered, the technicalities of most other things are soon revealed. The author anticipates where the difficulties lie and hopefully guides the reader through them.

Some definition of the term "modern electronics" may be in order. By "modern" is implied the aim directly towards semiconductor devices rather than via thermionic valves, which at the current rate of technological progress will soon be as old-fashioned as the gramophone horn. S.I. units (Système International d'Unités – the international system of units and symbols) are used throughout as no self-respecting book which claims to be modern could do otherwise, moreover it allows the series to offer the same tuition in any country where the English language is understood.

About the Series

Each book is a complete treatise of a particular branch of the subject and therefore can be used on its own with one proviso, that the later books do not duplicate material from their predecessors, thus a working knowledge of the subjects covered by the earlier books is assumed.

Book 1 contains all the fundamental theory necessary to lead to a full understanding of the simple electronic circuit and its main components.

Book 2 continues with Alternating Current Theory without which there can be no comprehension of speech, music, radio, television or even the electricity mains.

With this groundwork completed, *Book 3* follows on semiconductor technology, leading up to transistors and integrated circuits.

Books 1–3 therefore constitute a complete but very inexpensive basic electronic theory course of inestimable help in either career or hobby and it is expected that, provided that the need is established, other books will subsequently follow on specialized electronics subjects, all built on this earlier groundwork.

Appendices: In an effort to cater for readers of very many different levels of education, including those who unfortunately have received only the minimum, all study material which is educational but ancillary to electronics is included in appendices to which the main text refers directly or by a raised number, for example . . . orbits[(A8)] simply means that more information on orbits is contained in Appendix 8. Thus readers who are already conversant with the principles involved may decide not to digress from the main text. The appendices in this form also act as a convenient reference section and it is suggested that readers who are meeting most of the principles for the first time may wish to study the appendices first even if only to be aware of what help is contained in them.

Calculations: Serious study of any engineering subject must include a certain amount of calculation and because our intention is to learn electronics and not arithmetic we must use any artifice we can to avoid being sidetracked by laborious calculations. Help comes in several different forms, which include slide-rules, tables of logarithms and the electronic calculator. The latter is, without doubt, the most useful, it is capable of such speed and accuracy that calculations, far from being a burden, become hardly noticeable. Calculators cost so little nowadays that, with time, most homes are likely to have one. Certainly a calculator, even the simplest, would be a perfect accessory to this course and later on in the workshop, but it is not essential. Its relative cheapness has also more or less eclipsed the slide-rule which has the facility of speed but not accuracy. But for readers without either electronic calculator or slide-rule, logarithmic tables are quite adequate, and in fact skill in their use keeps one mindful of some of the processes which appear in more advanced work, especially with the unit of transmission, the decibel. Accordingly a complete appendix is included on how to use logarithms, but a set of tables has not been added because these are easily obtained at modest cost. Alternatively one may simply accept the calculations given in the examples in which case arithmetic aids can be dispensed with.

Practical Work: The benefits from some practical work cannot be over-emphasized, but exercises are not included because so many are already published by Bernard Babani (publishing) Ltd (see list at end of book) and also by the excellent electronics periodicals to be found at most newsagents.

1. THE NATURE OF ELECTRICITY

There was a time when a ray of light set off from a far distant star. It travelled through space for a million years at a speed which could encompass the Earth seven whole times in one second and it eventually arrived on Earth so that the star could be seen. A fact, not a fairy story but needing much belief and imagination, for how can we, who talk of ancient times, a mere two thousand years ago, ever appreciate a million years, five hundred times as long. Herein lies our first lesson in the fascinating study of electronics, that just as universe times and distances are so great as to be beyond the comprehension of ordinary man, so equally bewildering is the habitat of the electron which, in contrast is so utterly small and never seen. Yet we shall find that the universe and electrons have something in common.

But why bother? Surely electronics simply implies wiring bits and pieces together to produce calculators, computers, radios and televisions? Eventually yes, but there is much more and if we can learn some of the relatively simple laws of nature first and in our imaginations see electrons moving about according to these laws, then we shall be so much better equipped to *understand* the whole vast subject, whatever feature interests us.

1.1 The Electron

Consider first a very tiny piece of anything – water is convenient to start with because of its chemical simplicity. Imagine that we have a microscope so powerful that it can magnify more than one thousand million times (a laboratory microscope has a magnification of only about 1000). At this sort of magnification, which is equivalent to projecting a drop of water onto a screen stretching from the East to the West coast of the U.S.A., we should then be able to see the *molecules* of water as in Fig. 1.1. One molecule consists of an oxygen *atom* linked to two hydrogen atoms as shown. O is the chemical shorthand for oxygen and H for hydrogen, hence the well-known formula for water, H_2O. All the molecules are in motion around each other yet the whole drop stays complete because of their mutual attractions. The oxygen and hydrogen atoms are known as *elements*, that is, they cannot be resolved into anything more simple. Altogether there are one hundred or so elements and everything on Earth is made up either of these or of different combinations of them.

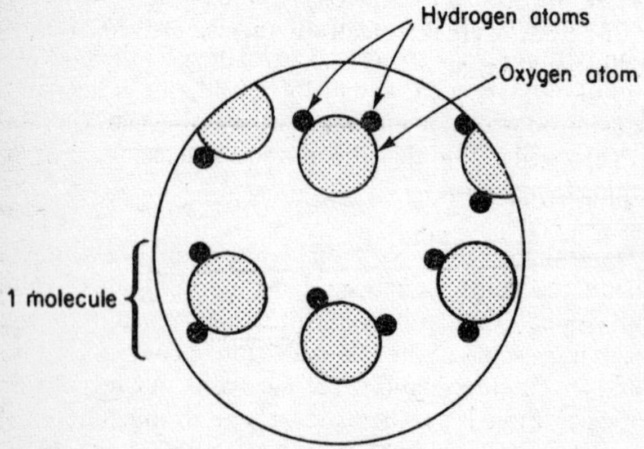

Fig. 1.1 Molecules of water.

But we have not yet reached the electron for which even more magnification is required to look right inside the atom. The hydrogen atom is the simplest and Fig. 1.2 shows pictorially its arrangement, that is, of a nucleus around which a single electron is in orbit(A8) just as in the universe, the Earth itself orbits the Sun.

The atom is not really as neat and tidy as shown, nor are its dimensions in the same relationship. Also much of its behaviour is still not completely understood, hence the earlier remarks on the need for imagination. Nevertheless we shall be concentrating more on the characteristics of the electron and sufficient is known of these to give us all the help we need.

Nearly all of the weight of the hydrogen atom is in its nucleus, it is some two thousand times the weight of the electron. The distance separating the electron from the nucleus is enormous when compared with the size of either of these, in fact most of the volume of the atom is space! To get this into some sort of perspective, think perhaps in terms of a football pitch, the ball, high in the air at the centre could represent the nucleus whereupon the orbits of the electrons would extend over the whole pitch although the electrons themselves would be smaller than golf-balls.

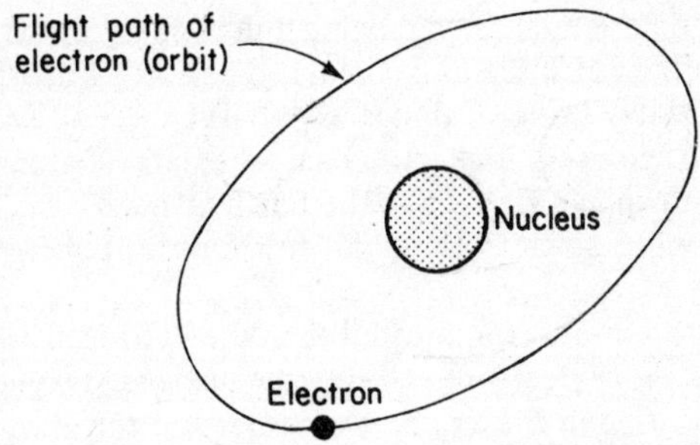

Fig. 1.2 Atom of hydrogen.

From what has been learned in Appendix 8, for the electron to continually orbit in space it must be moving at a certain velocity and there must be a force or pull attracting it to the nucleus. This brings us to another concept, that of *charges*. Many things in life we grow up with and accept without realising perhaps that we cannot really explain how they work, they are features of Nature itself. What we can do is to find the laws which relate to them, more how they act rather than why they do. The attraction due to gravity is an example. The charge of an electron is another, it is what we call electricity, we can calculate its weight and size and we can push it around to create any number of fantastic electronic devices, we can find out the laws which govern its effects, yet we cannot go further and explain what a charge is. So we accept that an electron is a little charge of electricity and delve no further.

Now gravity is always a force of attraction between two bodies but with charges we have two sorts called positive (shorthand, +) and negative (–). There is a most important but simple law to be learned and never forgotten: "Like charges repel, unlike charges attract." An easy reminder is through people, by imagining that male and female attract whereas male and male or female and female repel. This law allows us to explain the force keeping the electron in orbit round its nucleus for while there is a negative charge on the electron, there is a positive charge on the nucleus and being unlike charges, they attract. Just as the Earth seems to create some sort of invisible influence around it causing all things within its sphere to feel a force acting on them, so also electric charges create what in electronics is known as a *field*. But the Earth attracts everything, a charge only attracts an unlike (or opposite) charge.

All other atoms are more complicated than the hydrogen atom in that there are more electrons orbiting the nucleus. Oxygen has already been used in the example of water and the atom has eight electrons, never colliding because they have like charges which create repulsion when any one gets near to another.

An element we shall later be concerned with in transistors is silicon, which has 14 electrons and with even more is uranium with 92, so well-known in atomic energy projects. The complete atom is electrically neutral, it has no net positive or negative charge. This is arranged by the nucleus having a charge which is equal and opposite to the total charge of all the electrons. The nucleus itself contains two different types of particle, protons and neutrons. Neutrons have no charge and are of no concern to us in electronics, on the other hand each proton carries a positive charge of the same magnitude as the negative charge of the electron. The nucleus therefore contains the same number of protons as there are electrons in orbit, rather as in Fig. 1.3 for copper which has 29 electrons. (Remember that in this and in many diagrams which follow, we have to show a volume as best we can on a flat piece of paper.) The forces between these charges and within the nucleus are enormous compared with gravity. One has only to recall the devastation of a whole city by the explosion of a tiny atomic bomb to realise just how much energy lies within.

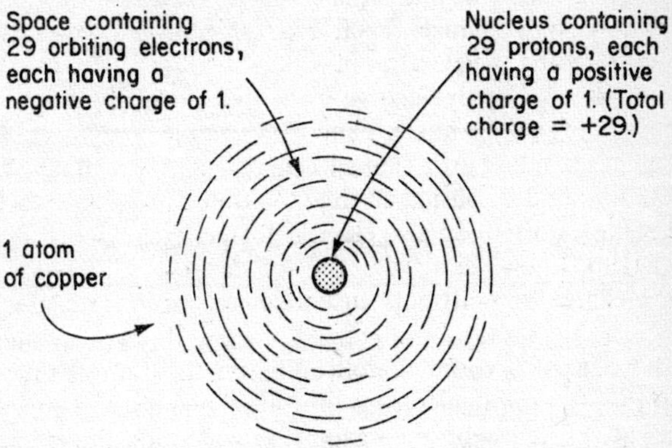

Fig. 1.3 Balance of charges in copper atom.

1.2 Electric Current

We are now in a better position to understand electronic (also called "electric") current flow. Electrons in the outer orbits of atoms experience less attraction to the nucleus than those nearer. In certain materials the electrons may be forced out of their orbits, for example, by the action of heat or the force due to a neighbouring charge, leaving them free to move among the other atoms. When these *mobile* electrons are made to flow in one direction, they constitute an *electronic current*. One or more free electrons at a point produce a negative charge, similarly one or more atoms which have lost electrons constitute a positive charge because an atom minus electrons has more positive protons than negative electrons, hence overall is positive.

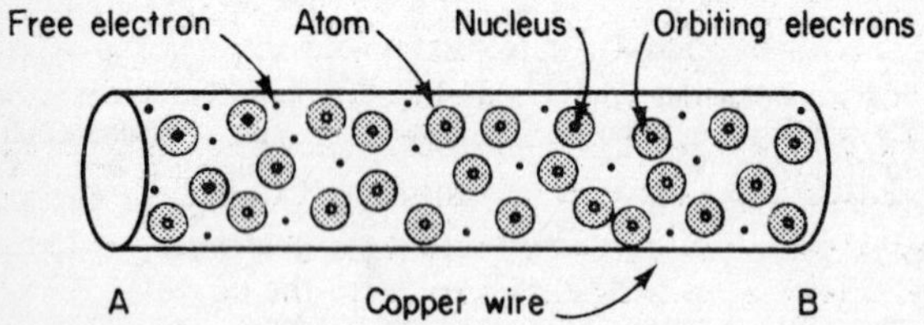

Fig. 1.4 Atoms and mobile electrons in a copper wire.

Fig. 1.4 illustrates how an electronic current flows in a conductor and to be able to add some numbers, let us assume that the conductor is an ordinary piece of copper wire about 2 mm[(A6)] in diameter and some 30 cm (the length of a desk ruler) long. Within this small piece of wire there could be as many as one hundred thousand million million million free electrons moving at random between the atoms. A number of this magnitude which in mathematical shorthand[(A2)] is written as 10^{23}, is beyond comprehension, this particular number need not be remembered, it merely serves to remind us how diagrammatic Fig. 1.4 is in being able to show only a few atoms when in fact a truly enormous number is involved.

Let us assume that a negative charge is applied to the end A of the wire and a positive charge to end B (this is what a battery or generator can do as we shall see later). The free electrons at A will be repelled by the external negative charge and at B free electrons will be attracted towards the positive charge, both effects causing an electron flow from left to right. Because electrons repel each other, there will be a movement of them in the same direction all along the wire, shifting from atom to atom, this constitutes current flow. The time taken for the electrons to be set in motion along a 30 cm length of wire is so small that for the present we may class it as instantaneous.

As an aid to the memory, one might imagine each atom nucleus to be a beehive and the electrons a swarm of bees around each hive. Most bees are very attached to their own hive and stay close to it, but a few are not so restrained and they buzz around at a greater distance, some even leave the vicinity and wander further afield, especially in warm weather. All our bees are aloof and antisocial, they keep their distance rather than meet up with one another. Connecting the battery is equivalent to other bees entering the gate at A, whereupon there will be a rush towards B to avoid the crowded condition brought about by the newcomers and some make their escape at B.

It is evident that the whole *conduction* process must depend on the availability of mobile electrons. Those substances, usually metals, which have large numbers are known as *conductors*; an electronic current flows through them easily and a well-known practical example is copper. Other materials have very few free electrons and these are known as *insulators* because it is very difficult to pass a current through them. In between there is a range of materials known as *semiconductors*, neither good conductors, nor insulators, but somewhere in between. These are of great importance in transistors which in fact are classed under the general heading of "semiconductor technology". The conductivity of a material is easily measured but to understand this we must study the measurement of current flow first.

1.2.1 Measurement of current

Electronic current in a wire can be likened to water flow in a pipe. The latter could be measured by the amount of water, say in gallons or litres[A6] flowing out of a tap per minute, that is, a *quantity* of water in a chosen *time*. Equally with electron flow we measure the number (or quantity) of electrons flowing past a point in a certain time and our unit of electronic current is known as the *ampere* (after the French physicist André Marie Ampère) and for one ampere the number of electrons passing is approximately 6.25×10^{18} per second(A2). In Fig. 1.4, for a current of one ampere flowing along the conductor from A to B, this is the number of electrons entering at A, the number leaving at B and the number passing any point along the wire, all within one second.

The ampere is a very important unit in electronics. We usually drop the term "electronic" or "electric" and simply refer to "current". The range of practical current values is very large, from the current flowing in a radio aerial which may be of the order of nano-amperes(A6) (one thousand, millionth of an ampere) to the thousands of amperes needed to start a train. In between there are, for example, the motor car headlamp which passes about 3 amperes and the 100-watt lightbulb with just over 400 milliamperes.

Of a little less practical importance is the measurement of the quantity of electricity itself (like the litre of water). Whereas with water we first take a quantity then define its rate of flow, we work backwards in electronics by defining as we have done, the rate of flow first and then the quantity from this. The unit of quantity of electricity (or more exactly, of electric charge) is called the *Coulomb* (after the French physicist Charles Augustin de Coulomb) and it is that quantity of electricity conveyed in one second by a current of one ampere and if for one ampere there are about 6.25×10^{18} electrons passing per second, then one coulomb of charge is shared by this number of electrons, i.e.

Quantity or charge of a single electron

$$= \frac{1 \text{ Coulomb}}{6.25 \times 10^{18}}$$

$= 1.6 \times 10^{-19}$ coulombs

(help is given in Appendix 2 in handling these figures).

Again, these incredibly large and small numbers are mainly for interest to emphasize that although electronics theory has to be based on mathematical calculation and experiment only, because we cannot see what happens, sufficient is known at present to explain all the phenomena about which we shall need to know.

The most important points to remember or visualize so far are that:–

(i) the electron carries, or rather, is, a negative charge which we measure in coulombs;
(ii) the charge creates a (negative) *field* around the electron which decreases in strength as the distance from the electron increases;
(iii) when another charged body enters this field there will be a mutual force of attraction if the new charge is positive but repulsion if negative;
(iv) current is simply a flow of electrons (*charge carriers*) in one direction and is measured by the number of electrons passing a given point in one second, the unit of current being the ampere.

It is shown in Chapter 5 how a meter for the measurement of current is constructed.

1.2.2 Factors influencing current flow

So far we have considered current flow using copper wire as an example simply because it happens to be a metal much used for this purpose, having an abundant supply of mobile electrons. We shall find however that metals differ in the ease with which their atoms give up electrons and with those having atoms which cling tightly there is obviously going to be more difficulty in creating current flow. It is important and of interest to have some conception of the mechanism involved, but always remembering that what follows is a very simplified theory, nevertheless useful in enabling us to appreciate just sufficient of the complex phenomena of atomic structure for our particular need.

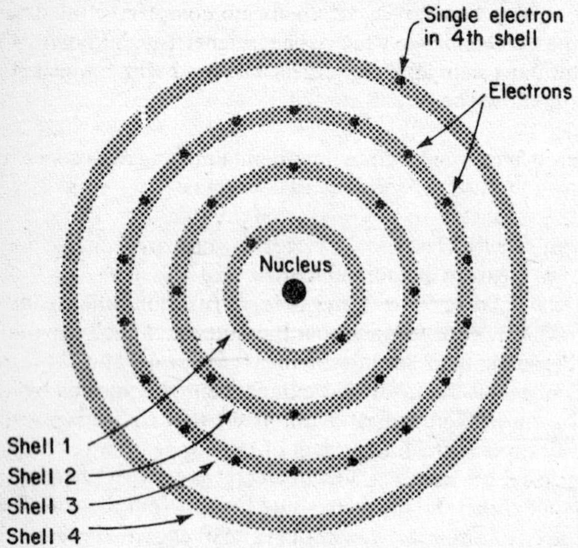

Fig. 1.5 Copper atom showing arrangement of electrons in shells.

In a rather general way, Fig. 1.3 shows a number of electrons orbiting their nucleus. Closer examination would show that the individual orbits are grouped at varying distances from the nucleus and the copper atom of Fig. 1.3 has been re-drawn in Fig. 1.5 to show this, again with the reminder that the figure represents a spherical volume rather than a flat circle. The electrons are constrained to orbits within *shells* as shown. Each element has what physicists call an atomic number which is in fact the number of protons in the atom nucleus, and also the same as the number of electrons orbiting the nucleus when all are present. There can be up to seven shells and the maximum number of electrons which any shell can accommodate is equal to twice the square of the shell number (see Fig. 1.5), for example:–

$$\text{Shell } 4 \ldots 4^2 = 16, \quad 2 \times 16 = 32$$

Up to atomic number 18 the shells are completely filled in order but for some elements having higher numbers things get a bit more complicated, some shells not being completely filled before the next is started.

Copper, however, which has an atomic number 29, has well-organized atoms, each shell is filled completely before electrons are allocated to the next, that is, Shell No. 1 has 2 electrons, Shell 2, 8, Shell 3, 18, leaving only one for Shell 4 as shown in Fig. 1.5. Remembering that the nucleus is positive and therefore exerting a force of attraction on its electrons, it is clear that those in the outermost shell experience the least force because of their greater distance. Also they are shielded to a certain extent from the nucleus by the electrons (like charges) in the inner shells. Thus the single electron in the fourth shell of the copper atom is easily disengaged from its orbit.

If we next consider the element having 30 electrons, which is zinc (again a well-organized atom), there must be two electrons in the outer, fourth shell. The attractive force on the outer electrons due to the nucleus is slightly greater in this case because it contains 30 protons instead of the 29 for copper, but the shielding is the same. Although there are twice as many outer electrons, they are less easily removed, the net result being that zinc contains fewer mobile electrons than copper and does not therefore *conduct* a current as easily. Electrically we say that zinc has a lower *conductance* to current flow than copper.

We cannot carry this reasoning very far however, for nickel (atomic number 28 with two electrons in the outer shell, No. 4) and iron (atomic number 26, again with two electrons in Shell 4), both have the same number of electrons in the outer shell as zinc, yet have lower conductances, hence we must at this stage accept that conductance is a complicated function of the number of electrons in the outer shell, the position of the latter relative to the nucleus and the shielding provided by the inner electrons. But we are not studying physics and considering that we have only discussed a few elements, not the myriads of compounds, all we need to remember is that:–

(i) each element has a different number of electrons from one to just over 100;
(ii) electrons occupy each shell in order but all shells do not necessarily have their full quota;
(iii) it is only the electrons in the outer shell which take part in the conduction process;
(iv) as a general rule, the more complete the outer shell is, the less the element will conduct electricity,

and this will suffice in helping us to recall more easily the conduction process and later the underlying physics of transistors.

Conductance is therefore a measure of the ease with which a current flows in a material. Peculiarly enough, in electronics we usually are more interested in the opposite, that is, the degree of opposition which is offered to current flow and this is called ***resistance***. Because all materials have resistance to current flow the concept is very important, we have discussed a unit for the measurement of current, the unit for resistance will be introduced in Chapter 3.

1.2.3 Heating effect of a current

Music is the pleasant outcome of vibration and provides us with an easy way of visualizing the way of life of the atom. Consider for example, a guitar string which when plucked vibrates to and fro from its rest position. Because this is the centre of the total travel of the vibrating string it is called the ***mean*** position. In the same way atoms vibrate about a mean position (see Fig. 1.6) but in this case the vibration is continuous, that is, unlike the guitar string the atoms are never at rest.

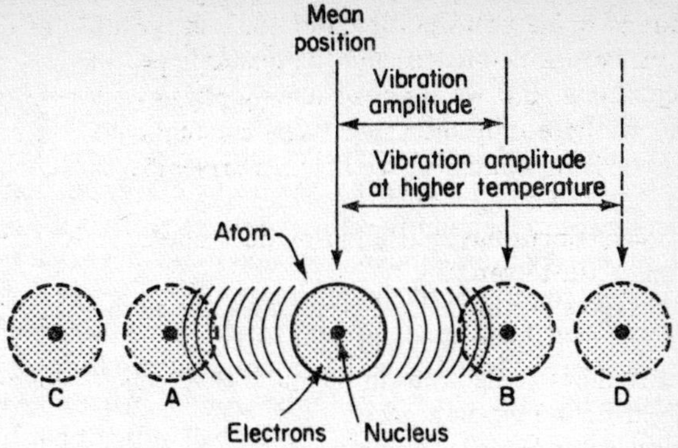

Fig. 1.6 Vibration of atom.
(The atom is considered as vibrating between A and B, and at higher temperature between C and D.)

In a solid material the atoms are arranged in fixed mean positions relative to each other, about which each one vibrates. The amplitude of vibration is proportional to the temperature of the material, at low temperatures it is small, at high temperatures, large (e.g. C–D, Fig. 1.6). Heating a solid material increases vibration amplitude, the atoms need to move apart with the result that the material expands. Heating still further can increase the amplitude so much that the atoms are forced out of their fixed positions and move around, the material has, in fact, melted so what was previously a solid has changed to a liquid capable of flowing.

Thus we must modify our mental picture of electrons merely darting to and fro between the atoms, sometimes joining up with one which has lost one or more outer electrons only to escape again later; it is more complicated because with atoms vibrating there is bound to be considerable interference between them and the electrons. This effect not only increases with temperature but itself produces heat which we know is generated by the passage of the current. Hence we must find a credible explanation, not perhaps good enough to satisfy a physicist but only to help us understand how things happen without perhaps being too far from the truth.

Although *energy* is a term in constant use, it is not one which is easily defined. Let us imagine a footballer with plenty of it. He kicks the ball and gives it some of his energy so that it accelerates from rest and flies through the air. That the ball possesses energy is amply demonstrated by the fact that not only does it travel against the resistance of the air but should it hit an unfortunate spectator he would certainly become painfully aware of it. The ball does not continue its journey, but having lost its energy, falls to the ground. What may not be appreciated is that the energy of the ball in flight has been transformed into heat on contact with the spectator. This amount of heat is small and would pass unnoticed.

In a similar way, because it is forced to move in a certain direction the electron can be likened to the football in that it receives energy which accelerates it. From the explanation above about vibrating atoms it is evident that many electrons will eventually collide with atoms, their energy being given up to the atoms as heat just as the football does to the spectator, again an extremely small amount but there are millions of collisions occurring. Thus the temperature of the material rises and effectively the current has produced heat within the conductor, the amount of heat clearly depending on the number of collisions.

The situation could be likened to a crowd of people, all milling around as at a fair; a group of hooligans mixes in and jostles everybody, there would be an increase in activity and plenty of heat generated!

Exploitation of the heating effect of a current is found in the home in the ordinary electric fire which normally runs at "red heat", or the lightbulb, with its filament heated to a much higher temperature. Continuing up the temperature scale, the heating effect is also used to protect electrical circuits against excessive current flow caused by fault conditions. A *fuse* is used which is a short piece of fine wire through which the current flows, the wire is capable of carrying the normal current without too much rise in temperature but should the current rise excessively, the increased heating within is sufficient to melt it ("blow the fuse") so that it breaks apart and disconnects the circuit.

1.2.4 Conventional current flow

Before progressing further, a word of warning must be injected about direction of current flow. The theory universally accepted now is that, as we have seen, electron (i.e. current) flow is from negative to positive. However, unfortunately, when earlier experimenters, who had not the benefit of electron theory to guide them, tried to decide on the direction of current flow, they chose the other way round and decided on positive to negative!

For a long time textbooks have explained in terms of both "conventional current flow" (+ to –) and electron flow (– to +) which has tended to add to the difficulties of students. Explanations in this series of books are all on the basis of electron theory, except where stated to the contrary.

1.2.5 Electrons at work

The ubiquitous electron can be found working in almost every walk of life and its field of activity continues to grow. A single modern piece of equipment is described below partly for interest but mainly to show how the principles outlined in this chapter are put into practice.

The Cathode-Ray (Television) Tube:

Fig. 1.7 shows the essential features of a "black and white" tube. At the left-hand end a heating filament raises the temperature of a *cathode*. The heat provides the mobile electrons with sufficient energy for many to escape from the surface, whereupon they would normally form a cloud round the cathode so repelling further emission. However, the first *anode* is made positive and this attracts the electrons and although some are collected on the anode a beam of them will shoot straight through the hole in the centre because of their high speed. There will be no interference from air atoms because the tube is evacuated (all air pumped out). The electrons are then accelerated further towards the more highly positive second anode. Again the beam shoots through the centre hole, along the tube to strike the coating of fluorescent material on the inside of the face. Where electrons hit the screen at high speed, the energy they give up causes the screen to glow. Focussing arrangements within the tube (not shown) ensure that the beam does not spread during its journey and produces only a tiny point of light.

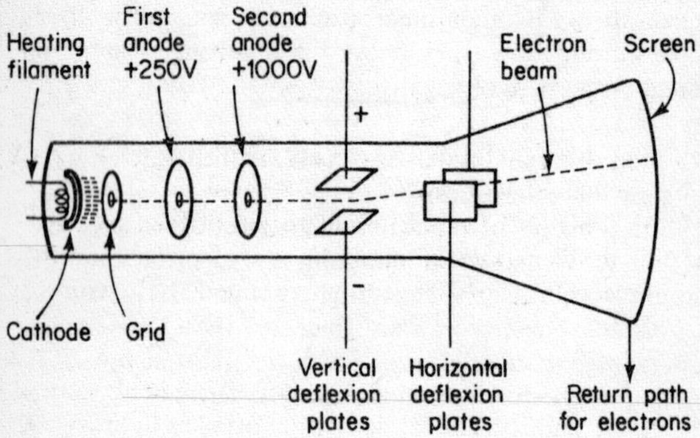

Fig. 1.7 Elements of a cathode-ray tube.

This is therefore a case of current flowing but not through a wire as we have so far considered, the electrons are projected through space.

The *grid* has a small negative potential applied to it and so because it will repel electrons according to the magnitude of its potential, it is used as a control. A high negative potential is sufficient to repel all electrons back to the cathode, a low one allows sufficient through to produce a bright spot on the screen.

Again demonstrating the law of like charges repelling and unlike attracting, there are two vertical deflexion plates. The plates themselves are horizontal and if, for example the upper plate is positive while the lower is negative, both will exert an upward bending force on the beam so the spot will move upwards on the screen and vice versa. Similarly the horizontal deflexion plates move the spot from side to side. Thus by swinging the potentials of both sets of plates, the spot may be made to trace out a complete rectangle as though it were ruling hundreds of lines on a page. It is made to do this so quickly that we only see a complete picture. As the spot moves over the screen it changes in brightness according to the picture voltage applied to the grid and so builds up the image we see. Colour tubes have the added complexity of three separate colour sections.

Thus the cathode-ray tube demonstrates:–

(i) heating certain materials can create the emission of large numbers of electrons;
(ii) the principle of repulsion and attraction enables the anodes to accelerate and guide the electron beam, the grid to change the spot brightness and the deflexion plates to make the beam trace out the picture area;
(iii) the energy of fast moving electrons is transferred into heat and light when they strike the screen.

2. THE SOURCES OF ELECTRICITY

In considering the movements of electrons we have discovered how current flows, how it produces heat as it does so and how materials differ in their abilities to conduct electricity. Next we examine the sources of electricity, that is, what conditions give rise to the charges which when connected across a wire cause a current to flow. The charges have been labelled + and –, however current will also flow if both charges have the same polarity, provided that there is a *difference* in charge. Under this condition current flows from the more to the less negative charge, for considering our piece of wire again, there will be a greater force repelling the mobile electrons in the wire at one end than the other. Alternatively it may be considered that the lesser charge, even though negative, is positive relative to the greater charge.

2.1 Chemical Action

This is illustrated best by going right back to the first chemical *voltaic* cell (after Alessandro Volta, an Italian physicist). Volta showed that zinc and copper plates separated by a cloth moistened with dilute sulphuric acid developed a difference in charge. Nowadays it is considered that the cell works from the *ionization* of the acid. As discussed in Chapter 1, a molecule of a substance is made up of two or more atoms of elements, the example being used of oxygen with hydrogen, producing a compound, water, completely different from both, which are gases. Similarly each sulphuric acid molecule is formed from two atoms of hydrogen, one of sulphur and four of oxygen, also described by the formula H_2SO_4. Sulphuric acid is well known as the *electrolyte* (the liquid in which the metal plates are immersed) in car batteries and accumulators. In this case the molecules divide when the acid is added to water, each into two hydrogen *particles* and one sulphate (SO_4) particle. The term *ion* is used for the particles because they are neither atoms nor compounds. The hydrogen atoms lose electrons to the sulphate to produce positively charged hydrogen ions while the sulphate ions are negatively charged because of their electron gain. The solution remains electrically neutral because the balance of charges has not been upset.

When zinc and copper strips are immersed in dilute sulphuric acid, positive ions just below the surface of each metal (atoms of the metal which have lost electrons) move into the acid to combine with the negative sulphate ions there. This leaves the metals negatively charged because of their surplus of electrons and the electrolyte with a surplus of positive ions. As the latter builds up it quickly repels further movement of positive metal ions. This is shown pictorially in Fig. 2.1.

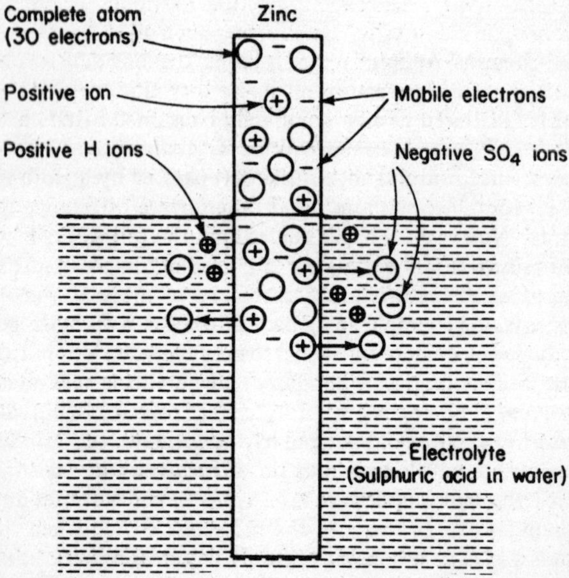

Fig. 2.1 Zinc in a dilute sulphuric acid electrolyte.

Zinc happens to have a much greater tendency to release positive ions to the electrolyte than copper hence the negative charge from the zinc to the solution is greater than that for the copper, thus the zinc is negative to the copper. Alternatively we could say that the copper is positive to the zinc, it does not matter which because all we are concerned with is that there is a difference in level of charge. Because this condition has the potential for causing current to flow, the zinc and copper are said to exhibit a *potential difference* which we will discuss in more detail in Chapter 3. However, from this reasoning we can certainly label the polarity of the cell as copper + and zinc –, to indicate the direction in which electrons will flow in an external circuit.

Thus when a wire is connected externally between the zinc and copper plates there will be electron flow along the wire tending to decrease the negative charge at the zinc and increase it at the copper. The chemical action continues by the zinc giving up more positive ions to the electrolyte to combine with the negative sulphate (SO_4) ions, forming zinc sulphate ($ZnSO_4$), the electrical result being that zinc maintains its surplus of electrons and is gradually consumed by the action.

This leaves a surplus of positive (H) ions in the electrolyte which combine with electrons at the surface of the copper to reform the ion into the original atom of hydrogen, the latter being given off as bubbles of gas forming on the copper.

The voltaic cell has many disadvantages as a provider of electricity, the main one being that the bubbles of hydrogen tend to stick on the copper plate, thus reducing considerably the area of copper exposed to the electrolyte (called *polarization*) so that the cell progressively fails. However, although ionization and the combination of ions and electrons may perhaps seem complicated, the explanation does demonstrate:

(i) chemical action in a specially designed cell can produce and maintain a difference in electric charge between two points;
(ii) when these two points are connected by a wire an

electronic current flows round the complete circuit consisting of the cell and wire;

(iii) the price paid for the production of this current is the gradual loss of the zinc by the action of the cell changing it into zinc sulphate.

In practice the Leclanché cell (after Georges Leclanché, a French chemist) is used, again with a zinc *electrode* (the metal in contact with the electrolyte) but instead of copper, a carbon rod is substituted. The electrolyte is ammonium chloride and the cell produces electricity by chemical action involving ionization as in the case of the voltaic cell. The problem of polarization by hydrogen bubbles is overcome by the use of a *depolarizer* which reacts chemically with the hydrogen to remove it as soon as it is formed.

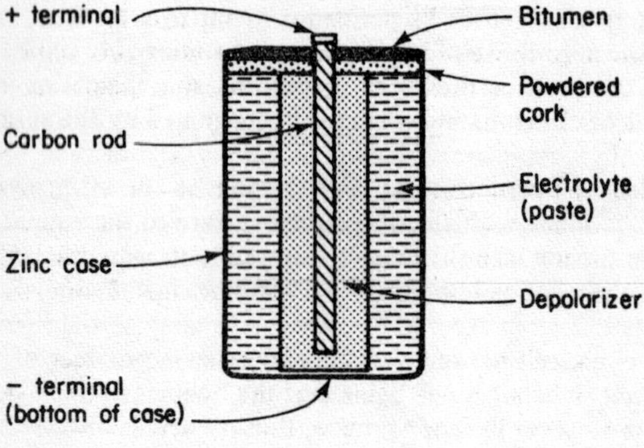

Fig. 2.2 Dry Leclanché cell.

Early Leclanché cells had a liquid electrolyte, modern ones use a moist paste, hence the term *dry cell*, most commonly found as torch batteries, transistor batteries, etc. Instead of a zinc rod or plate the cell has a zinc case which is its negative terminal, the carbon rod usually protrudes from the top of the cell and is the positive terminal. The depolarizer is a mixture containing manganese dioxide which reacts with the hydrogen gas produced at the surface of the carbon rod. The basic elements of the cell are shown in Fig. 2.2. Failure is frequently due to the electrolyte drying out rather than exhaustion of the zinc.

2.2 Magnetism

We are all familiar with magnetism if only from seeing a magnet pick up iron or steel objects, there obviously being some sort of invisible field emanating from the magnet which produces a force on certain objects. Magnetism is an inseparable part of electronics, even of the electron itself and so will be treated more fully later, the purpose in this section being only to show how it may be considered that electrons are caused to move in a wire by magnetism, thus setting up an electronic current. We now have to get to grips with a rather more difficult conception about electrons which is that they spin on their own axes at the same time as moving in orbits. Fortunately we have the Earth as an example for not only does it orbit the Sun once per year, but spins on its own axis, completing one revolution in 24 hours, so giving us day and night. It is now considered that magnetism arises from moving electric charges, thus each electron not only has its electric field around it but also a magnetic one. Just as there are two types of electric charge obeying the rules of like repelling and unlike attracting so also magnetic *poles* are of two different types, again obeying the same law. Thus when a magnetic field is made to move across a wire (or equally, a wire is moved through a magnetic field) so that this field and those tiny ones due to the electrons interact, the electrons are moved out of their orbits and caused to accelerate in one directon. This is of course an incomplete explanation but again, it is a useful basis on which to create a mental picture.

An electrical generator is simply a machine which either rotates wires through a magnetic field or rotates a magnetic field so that it cuts stationary wires, either method compelling the electrons in the wires to move along and through the external circuit.

2.3 Friction

Friction is the resistance a body or material meets when it is rubbed over another. During this action electrons are pulled out of their orbits at the surface of one of the materials and acquired by the other. The first material becomes positively charged through its loss of electrons while the second exhibits a negative charge by its gain. Many pairs of materials produce this *static electricity* (the electricity remains at rest on the material after the rubbing has ceased), for example, glass may be rubbed with silk, the glass acquiring a positive charge with the silk negative. Amber was also a material used in early experiments and from its Greek name "elektron" is derived the title of our subject.

Lightning is an unmistakable example of static electricity, being built up by friction on clouds as they move through the air, the charge having opposite polarity to that on the earth beneath. The potential becomes so high that the air breaks down by ionization between the clouds and the earth and a huge current flows through the ionized path, usually of many thousands of amperes and sufficient to produce a flash of light so well known to all.

Static electricity obeys the rule that connecting two different charges together causes a current to flow. Whereas with chemical action the current is maintained, with static charges it ceases when they have equalized.

2.4 Heat

Consider two different metals in contact, one having a large number of mobile electrons, the other only a few. Because the mobile electrons are darting about in all directions it is inevitable that more will cross the junction from the metal with the plentiful supply into the other metal than will cross in the opposite direction. Thus the metal with plenty will become positively charged by virtue of its loss of electrons while the other will become negative. The difference in charge between the two metals gives rise to what is termed *contact potential.*

The speed at which the electrons move increases with temperature, therefore at higher temperatures more electrons pass across the junction and increase the contact potential. Accordingly a heated junction will maintain a difference in charge even when the ends of the metals or wires are connected together to allow a current to flow. The device is known as a *thermocouple*. Because the contact potential is small it has little use as a generator of electricity but has many applications as a thermometer for by measuring the current flowing round a thermocouple, its temperature can be calculated.

2.5 Light

Light is a subject which physicists continue to explore in an attempt to get a clearer picture of how it functions and how we see. Hence we will have to be satisfied with a very simple theory of how light produces an electric charge and naturally it must fit in with the principles already found. A *selenium cell* is known to produce a tiny current in an external circuit

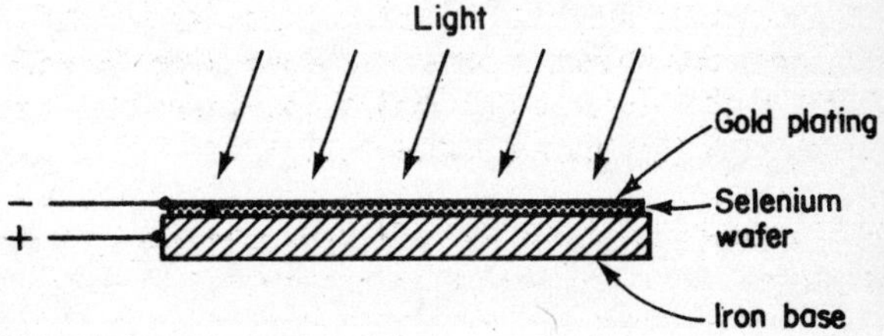

Fig. 2.3 Selenium cell.

when light falls on it, a typical cell consisting of a wafer of selenium sandwiched between gold plating and an iron base plate as shown in Fig. 2.3. The gold plating is thin enough to allow light to pass through to the selenium. When it does so, the energy of light (light is a form of energy just as heat is) strikes the atoms of the selenium giving some electrons sufficient energy to escape from their parent atoms and become mobile. As with a thermocouple, because of the increased availability of electrons in the selenium, some cross into the gold plating, so creating a difference in charge across the selenium/gold junction.

2.6 Pressure

Pressure can also be used to produce a difference in charge but in common with friction, heat and light, the magnitude is relatively small. Only a few substances exhibit the required effect, particularly quartz and Rochelle salt. These operate from the fact that when mechanical stress is applied to certain crystals, electrons are forced away from one surface towards another. The former therefore exhibits a positive charge relative to the latter and the difference in charge is sufficient to cause a tiny current to flow in an external circuit. Typical uses are in *crystal* microphones and pick-ups where in the first case the mechanical stress is provided by pressure from a sound diaphragm and in the second case directly from a needle which is caused to vibrate in the record groove.

3. THE SIMPLE ELECTRIC CIRCUIT

Chapter 2 has provided a brief resumé of the activities of the electron in the production of electricity, brief because the intention is no more than to show some practical effects of the theoretical considerations of Chapter 1 but thereby to increase our perception of the intricacies of the atomic world. By so doing we shall find the study of electronics more rewarding through a deeper understanding of how things happen even though we have only our mental pictures to guide us about our invisible subject. But we can see, hear and feel the effects of unbelievably large numbers of them when they are controlled to do what we desire and so in this chapter we move on from study of a few infinitesimally small charges to the combined handiwork of millions and millions of them together.

3.1 Ohm and His Law

The water analogy of Section 1.2.1 has compared water flow with current flow; we can take this one step further and liken the water pipe diameter to the conductance of a wire for the greater the pipe diameter or conductance, the greater the water or current flow will be. The analogy is not complete however until we include pressure because it is obvious that water flow along or out of a pipe is also affected by the water pressure. The equivalent to water pressure in electricity is the difference between the two charges, thus we have yet to determine what is the effect of the amount of difference in charge and to find some unit to express it, consisting as it does merely of a superior number of electrons at one point relative to that at another.

The term "potential" was introduced in Section 2.1 and that electricity has potential is in no doubt. Long before our time there have been thunderstorms in space and electric eels in water. These certainly had potential, to shock or even kill and throughout the ages people have known and respected them, with fears greater than their understanding. However, around the eighteen-hundreds scientists at last began to get to grips with conjecture through properly conducted experiments and we remember many of these famous people by naming our electrical units after them. We now meet Georg Simon Ohm, a German physicist who formulated his famous law which must be engraved on every technician and experimenter's heart.

With devices for measuring electricity becoming available, Professor Ohm studied how, what was therefore called, an electric potential, caused current to flow through a wire and he demonstrated that the magnitude of the current was directly proportional to the value of the potential applied across the wire. This means, for example, that if the potential is doubled the current doubles and so on. Mathematically this is written:

$$I \propto V$$

where I is the letter used to represent current and V is used to represent the potential difference in *volts* (after Volta).

Now we have already established that the current is also proportional to the conductance of a material, i.e.

$$I \propto G$$

where G is the letter used for conductance. Therefore

$$I \propto V \text{ and } G.$$

Ohm did not have the benefit of electron theory to guide him as we have, so his law does not talk in terms of conductance but of resistance as mentioned earlier. These are opposites, that is, a material having high resistance has low conductance and vice versa, in fact, mathematically it is said that one is the reciprocal(A4.1.3) of the other or

$$R = \frac{1}{G} \quad \text{and} \quad G = \frac{1}{R}$$

where R is the letter used to represent resistance.

Ohm defined the resistance of a material as the ratio of the potential (V) to the current (I) it produces and when a potential of 1 volt gives rise to a current of 1 ampere then by his law, the value of the resistance equals one and this was called the *ohm*. The general statement of Ohm's law is therefore

$$I = \frac{V}{R}$$

where I = current in amperes, V = potential difference in volts, R = resistance in ohms, and we should need no aids to memorize this because from what we have learned so far it is obvious that the higher the potential difference, the greater will be the pressure on the electrons to flow. Also the higher the resistance, the lower will be the current because of the greater reluctance of the atoms to give up electrons and vice versa. As a final comparison with water from the tap, greater pressure causes more water flow, greater resistance (smaller bore pipes or tap turned partly off), lower water flow.

By simple manipulation(A4.2)

$$V = I \times R \quad \text{and} \quad R = \frac{V}{I},$$

thus given any two we can easily determine the third.

Practical voltages vary over a very wide range, some indication of which is given in Fig. 3.1. In written work or drawings, amperes, volts and ohms are represented by A, V and Ω (the Greek capital letter Omega) respectively. These are to be found with their most commonly used multiples and sub-

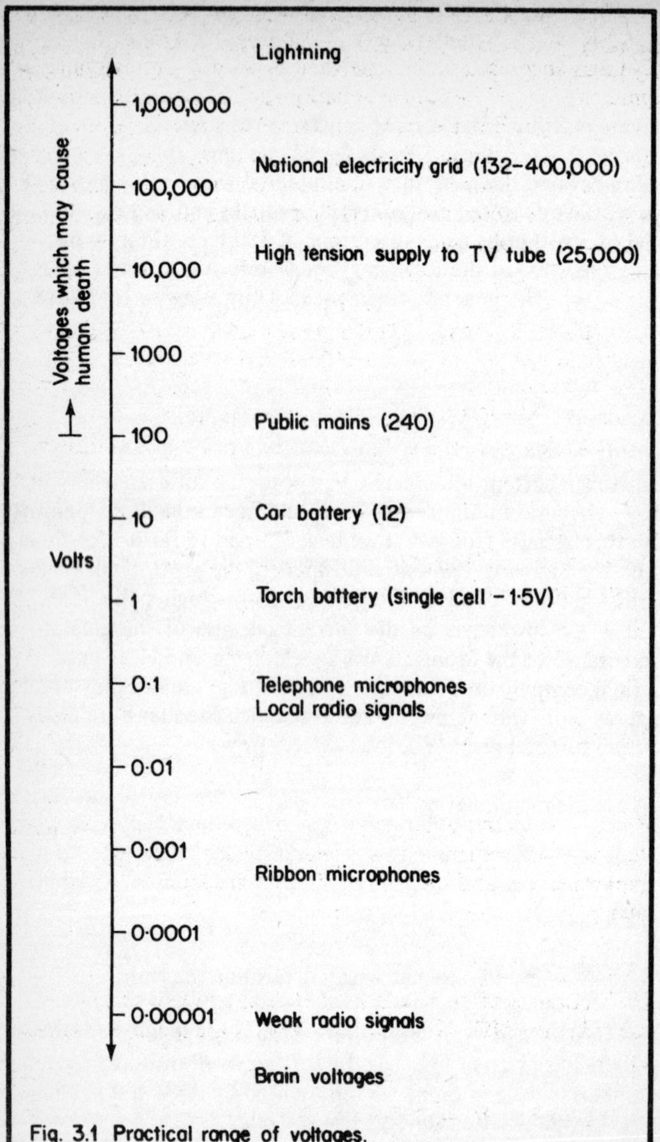

Fig. 3.1 Practical range of voltages.

multiples in Appendix 6, as are all other units to be met later.

Conductance used to be measured in an easy-to-remember unit, the "mho" but the internationally agreed unit is now the Siemen (after Ernst Werner von Siemens, a European inventor) and if a material has a resistance of one ohm, it has a conductance of one siemen (1 S – capital S so as not to be confused with time in seconds, denoted by a small s). It follows therefore for example, since

$$R = \frac{1}{G} \quad \text{and} \quad G = \frac{1}{R},$$

a resistance of 10 Ω is equivalent to a conductance of 0.1 S; a conductance of 50 S is equivalent to a resistance of 1/50 = 0.02 Ω.

EXAMPLE:
An electric fire element has a resistance of 50 Ω. What current flows if the mains voltage is 250?

$$I = ? \qquad V = 250\ \text{V} \qquad R = 50\ \Omega$$

$$\text{Current flowing, } I = \frac{V}{R} = \frac{250}{50} = \underline{5\ \text{A.}}$$

(Note – wires and other materials change their resistance when they heat up, we ignore this effect at present although it is an important one and therefore will be considered at length later.)

EXAMPLE:
The resistance of the insulation between two wires in an electric cable is 1 MΩ. What current flows if the mains voltage is 250 V?

$$I = ? \qquad V = 250\ \text{volts}$$

$$R = 1\ M\Omega = 1{,}000{,}000\ \Omega\ (10^6\ \Omega)$$

$$\text{Current flowing, } I = \frac{250}{1{,}000{,}000} = \underline{0.00025\ A}.$$

(Remember that in the formula all units used must be in amperes, volts and ohms.)

This is not expressing the current in a very practical way, so we might, for example, bring 0.00025 A to milliamperes (mA) by multiplying by 1000 (there are 1000 mA in 1 A). Thus

$$0.00025\ A = 0.00025 \times 1000\ mA = \underline{0.25\ mA}$$

or even to microamperes by multiplying by 10^6. Thus

$$0.00025\ A = 0.00025 \times 10^6\ \mu A = \underline{250\ \mu A}.$$

With a little more experience we would have recognized this from the beginning and converted to μA immediately. Thus

$$\text{Current flowing, } I = \frac{250\ A}{10^6} = \frac{250}{10^6} \times 10^6\ \mu A = \underline{250\ \mu A}$$

or using exponents directly as in Appendix 2:

$$I = 250 \times 10^{-6}\ A = 250 \times 10^{-6} \times 10^6\ \mu A = \underline{250\ \mu A}$$

(since $\frac{1}{10^6} = 10^{-6}$ and $10^{-6} \times 10^6 = 10^{-6+6} = 10^0 = 1$).

EXAMPLE:
A 12 V battery passes a current of 200 mA through a coil of wire. What is the resistance of the wire?

$$I = 200\ mA = \frac{200\ A}{1000} = 0.2\ A$$

$V = 12\ V \qquad R = ?$

Resistance of wire,

$$R = \frac{V}{I} = \frac{12}{0.2} = \underline{60\ \Omega}.$$

EXAMPLE:
An electric bell has a resistance of 30 Ω and carries a current of 100 mA. On what supply voltage should it be used?

$$I = 100\ mA = \frac{100}{1000} = 0.1\ A$$

$$V = ? \qquad R = 30\ \Omega$$

Supply voltage, $V = I \times R = 0.1 \times 30 = \underline{3\ V}$

The next two examples are slightly more difficult as they involve not only Ohm's Law, but also considerations from Chapter 1.

EXAMPLE:
A potential difference of 6 V is maintained for one minute across a wire having a resistance of 2 Ω. Calculate the total charge transferred.

From Section 1.2.1, the unit of charge transferred, the Coulomb (C) constitutes a current of 1A flowing for 1 second (s). By Ohm's Law

$$\text{Current in wire} = \frac{V}{R} = \frac{6}{2} = 3\ A$$

Charge transferred when 3 A flows for 1 s = 3 C

∴ Charge transferred when 3 A flows for 60 s =

$$3 \times 60 = \underline{180\ C}$$

EXAMPLE:
A wire has a conductance of 0.2 S and a charge of 40 C is transferred through it in 5 seconds. Calculate the current in the wire and the potential difference across it.

If 40 C are transferred in 5 s, then

$$\frac{40}{5} = 8 \text{ C are transferred in 1 s, i.e. current} = \underline{8 \text{ A}}$$

By Ohm's Law

$$I = \frac{V}{R} \quad \text{and since} \quad G = \frac{1}{R}$$

(note G = conductance in Siemens)

$$I = V \times \frac{1}{R} = V \times G$$

$$\therefore V = \frac{I}{G}$$

Now $I = 8$ A $V = ?$ $G = 0.2$ S

∴ Potential difference across wire,

$$V = \frac{8}{0.2} = \underline{40 \text{ V}}$$

3.1.1 Power
(This section is complete but for those readers who are unsure of the technical aspects of force, work, energy and power, Appendix 9 treats these subjects in more depth.)

In our water analogy we considered water flowing in a pipe to have similarities to current flowing along a wire. Just as a pump or gravity is needed to do the work necessary to make water flow, so also work has to be done to move a quantity of charge, that is, to make current flow. In the electrical case the work is done by the force due to the potential difference V, in moving a quantity of electricity (or charge) Q between the two points, this having to be carried out against the electrical resistance between those points.

The internationally agreed unit of work is the joule[A9] and the work done in moving Q coulombs through V volts is expressed by

$$W(\text{joules}) = Q(\text{coulombs}) \times V(\text{volts})$$

Now power (P) is the rate of doing work, therefore

$$\text{Power} = \frac{\text{Work done}}{\text{time}} = \frac{Q \times V}{t}$$

and with Q in coulombs, V in volts and t in seconds, we arrive at the electrical unit of power, the *watt* (after James Watt, the English inventor):

$$P(\text{watts}) = \frac{Q(\text{coulombs}) \times V(\text{volts})}{t(\text{seconds})}$$

and since Q x V is the number of joules of work done,

$$1 \text{ watt} = 1 \text{ joule per second}$$

But 1 coulomb of electric charge flowing past a point in 1 second is equal to a current of 1 ampere, i.e.

$$I(\text{amperes}) = \frac{Q(\text{coulombs})}{t(\text{seconds})}$$

$$\therefore \text{since } P = \frac{QV}{t} = V \times \frac{Q}{t}$$

$$\therefore P = V \times I$$

Now, from Ohm's Law $I = \frac{V}{R}$

$$\therefore P = V \times \frac{V}{R} = \frac{V^2}{R}$$

Again, from Ohm's Law $V = IR$

$$\therefore P = I^2 R$$

Thus, summing up:

$$P = VI \quad \text{or} \quad \frac{V^2}{R} \quad \text{or} \quad I^2 R$$

where P is the power dissipated in watts, V, I and R are in volts, amperes and ohms.

These formulae are important. Electrical power can be employed to do many forms of work such as turning the shaft of an electric motor or producing heat. The heat may be useful as in the case of an electric fire, but conversely in electronics it can be damaging to small components if allowed to become excessive.

EXAMPLE:
A wire having a resistance of 10 Ω carries a current of 6 A. What power is dissipated in the wire?

$$I = 6\,A \qquad R = 10\,\Omega \qquad P = ?$$

Then $P = I^2 R = 6^2 \times 10 = 36 \times 10 = \underline{360 \text{ watts}}$

EXAMPLE:
A loudspeaker has a resistance of 4 Ω and an amplifier develops 0.2 V across it. How much power is dissipated?

$$V = 0.2\,V \qquad R = 4\,\Omega \qquad P = ?$$

$$\text{Then } P = \frac{V^2}{R} = \frac{0.2^2}{4} = \frac{0.04}{4} = 0.01 \text{ watts}$$

$$= 0.01 \times 1000\ \text{mW}^{(A6)} = \underline{10\ \text{mW}}$$

EXAMPLE:
What current flows through a 100 watt electric lightbulb if the mains voltage is 250? What is the resistance of the lamp?

$$P = 100\,W \qquad V = 250\,V \qquad I = ? \qquad R = ?$$

$$\text{Then } P = V \times I$$

$$\therefore I = \frac{P}{V} = \frac{100}{250} A = \frac{100}{250} \times 1000\ \text{mA} = \underline{400\ \text{mA}}$$

$$\text{Also } P = \frac{V^2}{R}$$

$$\therefore RP = V^2$$

$$\therefore R = \frac{V^2}{P} = \frac{250^2}{100} = \frac{25\not{0} \times 25\not{0}}{1\not{0}\not{0}} = \underline{625\ \Omega}$$

which, because we have already calculated I, could also have been obtained from

$$R = \frac{V}{I} = \frac{250}{0.4} = \underline{625\ \Omega}$$

3.2 The Circuit

So far current flow has been discussed mainly in terms of a wire having a potential difference (p.d.) maintained across its ends. Chapter 2 has shown that all sources of electricity have two "poles" and if one of these has a greater negative charge than the other, a current can be caused to flow through the wire. Connecting a source across a wire forms the simplest electrical circuit, the term "circuit" indicating that the pathway for the current is continuous, that is, it doubles back on itself, hence all sources must have a path through them with the same electron flow through the source as in the external circuit. As an example, suppose a 3 V torch battery lights a bulb and the current through the bulb is 0.1 A, then there is a current through the battery itself which must also be 0.1 A. The torch is used as an example of how a circuit diagram is set out in the next section.

3.2.1 Drawing the circuit

Electrically a hand-torch is represented by a circuit diagram as developed in Fig. 3.2(i) and (ii). Electrons from the battery flow from its zinc case through the spring in the bottom of the torch and metal case, to the switch which is moved upwards to make contact with the reflector. The lamp bulb is screwed into the reflector and the circuit continues through the lamp filament to the pip at the bottom of the shank and thence to the positive pole of the battery. Within the battery the chemical action is such that its terminal p.d. is maintained. The current through the lamp filament causes it to become white hot and produce light. Electrons, being only electrical charges and thus having practically no weight, flow immediately the switch makes contact, equally when the circuit is broken by the switch they instantly cease their controlled flow and resume random movements among the atoms.

In the circuit diagram the switch is shown in its "normal" position, which in the case of a pocket torch, is open. Note from Fig. 3.2(ii) that when the switch is closed, there is a complete electrical circuit, that is, a path can be traced right round back to the starting point. Other practical arrangements for a pocket torch may be used but they will always result in the same circuit.

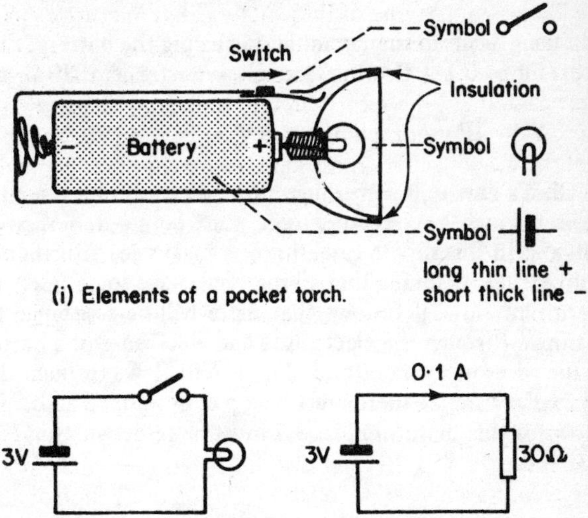

(i) Elements of a pocket torch.

(ii) Circuit diagram of (i).

(iii) Representation of working circuit.

Fig. 3.2 Circuit diagrams of pocket torch.

3.3 Electromotive Force

By Ohm's Law, with a 3 V battery and a current of 0.1 A flowing, the resistance of the lamp, R is V/I = 3/0.1 = 30 Ω, thus we could also represent the working circuit as at (iii) in Fig. 3.2. Note the general symbol for a resistance. The value is shown adjacent to the symbol. As further symbols are introduced, they will not be discussed in the text as for convenience all used are shown in Appendix 7.

Fig. 3.2(iii) is theoretically correct because the circuit current, voltage and resistance values obey Ohm's Law. However, if we were to measure the battery voltage with and without the *load* connected, we should find a slightly higher value in the second case.

Referring again to the circuit and ignoring the battery, there is a current of 0.1 A flowing through a resistance of 30 Ω and

$$V = IR = 0.1 \times 30 = 3\ V.$$

So Ohm's Law tells something more, which is that if a current flows through a resistance, there must be a p.d. across it (obvious in this simple case, but not always so). Furthermore, from earlier reasoning that all materials have some resistance to current flow, it follows that there will be resistance for example, through the electrolyte and electrodes of a battery or the wires in a generator and if current flows through this ***internal resistance*** there must be a p.d. developed across it. To allow for this therefore Fig. 3.2 must be re-drawn as in Fig. 3.3.

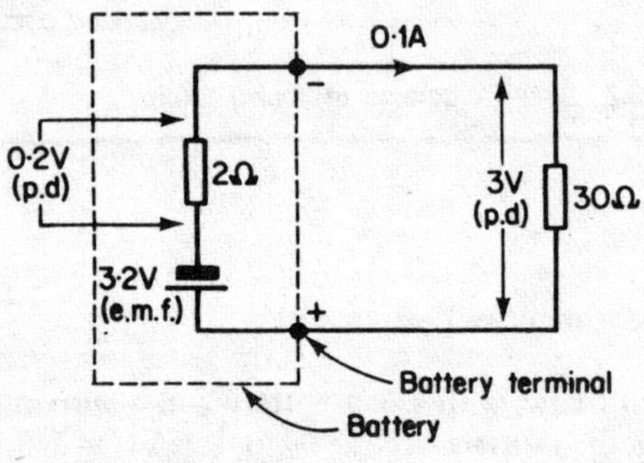

Fig. 3.3 Battery with internal resistance.

Suppose that the path through the battery has a resistance of 2 Ω. Then by Ohm's Law:

p.d. across 2 Ω resistance with 0.1 A flowing

$= 0.1 \times 2 = 0.2$ V

and if at the terminals of the battery there is a p.d. of 3 V, the chemical action of the battery is actually generating a force equal to 3.2 V. This is known as the *electromotive force*, usually abbreviated to "e.m.f." and also measured in volts. Thus in Fig. 3.3 the battery has an e.m.f. of 3.2 V and a p.d. when connected to a resistance of 30 Ω of 3.0 V.

The rule for the simple circuit is therefore that the sum of the p.d.s is equal to the e.m.f. as shown in Fig. 3.3. This also helps to explain how a dry cell fails electrically for as its internal resistance increases due to drying up of the paste electrolyte or failure of the zinc electrode, more voltage is *dropped* (and therefore lost) across this resistance, leaving less available at the terminals to drive a current through the external circuit.

3.3.1 Resistances in series

The term "simple electric circuit" has been chosen to indicate one in which the current is the same at any point, the components of the circuit are then said to be in *series*, that is, the current meets them in succession. Considering the resistance parts of Fig. 3.3 we arrive at Fig. 3.4(i) and recall that by

Ohm's Law:

voltage across resistance of 2 Ω = $0.1 \times 2 = 0.2$ V and
voltage across resistance of 30 Ω = $0.1 \times 30 = 3.0$ V

and the voltage applied to the two resistances in series is the sum of these, i.e. 3.2V, demonstrating two very important facts about the series circuit:

(i) the current is the same throughout;
(ii) using the general term *voltage* for both p.d. and e.m.f., the voltage across both resistances is equal to the sum of their individual voltages.

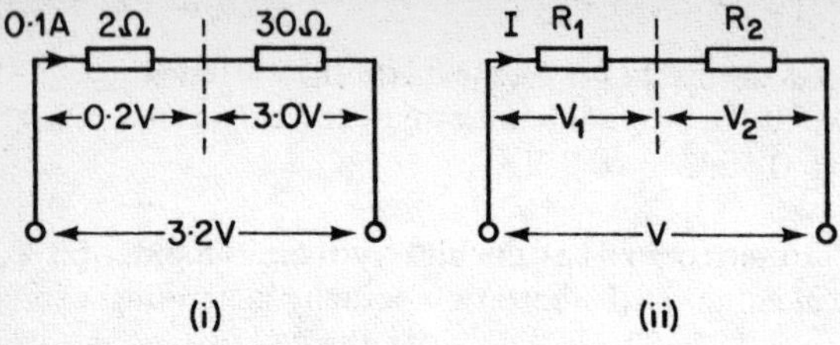

Fig. 3.4 Two resistances in series.

The circuit of Fig. 3.4(ii) is a general one where any values may be used as long as Ohm's Law applies, and R_1 (ohms) represents one resistance, V_1 (volts) the voltage across R_1, similarly with R_2; the current is I (amperes) and the voltage applied to the combination is V (volts).

Let R (ohms) represent the net resistance of R_1 and R_2 (i.e. the total resistance of the circuit). Then

$$V_1 = IR_1 \qquad V_2 = IR_2$$

$$V = V_1 + V_2 = IR_1 + IR_2 = I(R_1 + R_2)$$

Hence $\dfrac{V}{I} = (R_1 + R_2)$

But $\dfrac{V}{I}$ is also equal to R,

$$\therefore R = R_1 + R_2$$

that is, the total resistance of R_1 and R_2 in series is equal to $(R_1 + R_2)$. Using the same method of proof for three or more resistances in series will show that the total resistance of any number of individual resistances in series is equal to their sum, i.e.

$$R = R_1 + R_2 + R_3 + \ldots$$

EXAMPLE:
In the circuit of Fig.3.3 what will be the current through the lamp when the internal resistance of the battery has risen to 10 Ω?

It is often instructive to draw the equivalent circuit, e.g. as in Fig. 3.5.

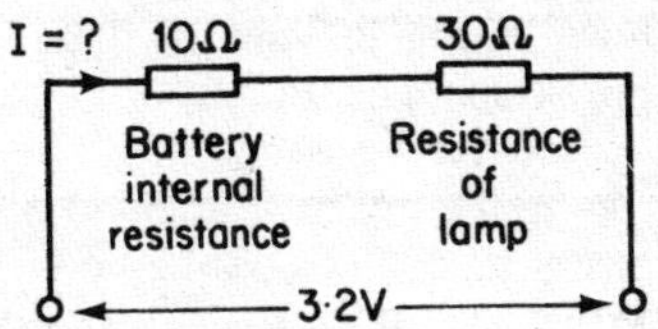

Fig. 3.5 Calculation of current.

$$\text{Total resistance of circuit} = 10 + 30 = 40\ \Omega.$$

Then, by Ohm's Law

$$\text{current through lamp} = I = \frac{3.2}{40} A = 0.08\ A$$

$$= 0.08 \times 1000\ mA = \underline{80\ mA}$$

EXAMPLE:
Three resistances are connected in series, 200 Ω, 800 Ω and 1.5 kΩ. They are connected across a 100 V supply. What is the current and the p.d. across each resistance?

Let resistance of combination = R Ω (see Fig. 3.6). Then,

$$R = 200 + 800 + 1500 = 2500\ \Omega$$

and by Ohm's Law:

$$I = \frac{100}{R} = \frac{100}{2500} = \underline{0.04\ A}$$

p.d. across 200 Ω resistance = 0.04 x 200 = 8 V

p.d. across 800 Ω resistance = 0.04 x 800 = 32 V

p.d. across 1.5 kΩ resistance = 0.04 x 1500 = 60 V

These p.d.s are shown in Fig. 3.6 and it will be seen that they add up to the supply value, i.e. 100 V.

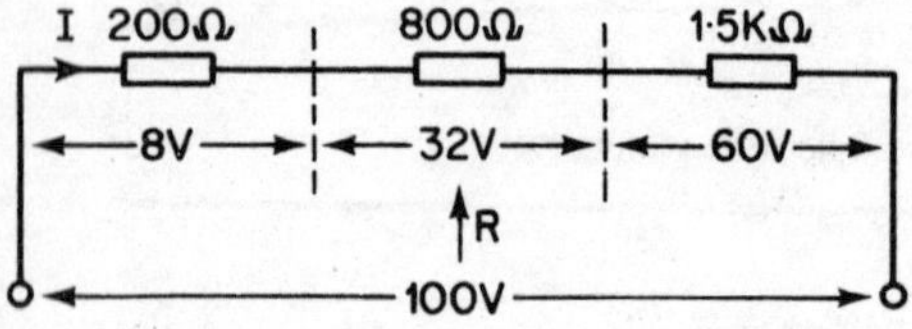

Fig. 3.6 P.D's across series resistances.

EXAMPLE:
A panel lamp works at a resistance of 60 Ω and requires a current of 100 mA. On what voltage should it be used?

$$V = IR = \frac{100}{1000} \times 60 = \underline{6\ V}$$

(remember to bring the 100 mA to amperes).

EXAMPLE:
If there is only a 9 V supply available, what value series resistance could be used with the lamp in the above example for correct working?

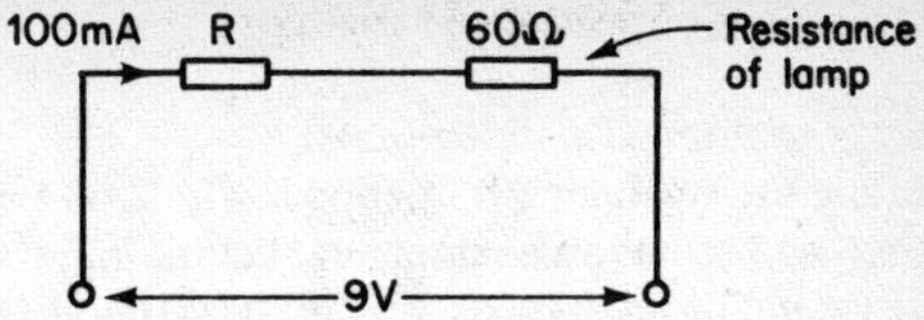

Fig. 3.7 Dropping resistance with lamp.

The circuit diagram is as in Fig. 3.7. Let the required resistance = R Ω. Then total resistance

$$= (R + 60)\Omega$$

and total resistance

$$= \frac{V}{I} = \frac{9}{0.1} = 90\ \Omega$$

$$\therefore\ R + 60 = 90$$

$$\therefore\ R = 90 - 60 = \underline{30\ \Omega}$$

Check as follows:
p.d. across R = 0.1 x 30 = 3V, i.e. if 3 V is "dropped" across R, there remains 6 V across the lamp, which is the value required.

3.4 Resistance

Having established the concept of resistance and how it forms part of the simple circuit, it is essential that we pause to examine how resistance varies with other parameters. This is not only to develop a better understanding of resistance itself, but also because it appears in an unwanted form in many circuits and components, moreover at times it changes in value, thus setting a trap for the unwary. Good design and measurement therefore cater for the effects of unwanted resistance and of resistance variations.

3.4.1 Variation with dimensions

Recalling the small length of copper wire 2 mm in diameter, 30 cm long used as an example in Section 1.2, it will have a certain value of resistance (actually 0.0016464 Ω or 1646.4μΩ as will be shown in an example later). Doubling or trebling the length of wire to 60 or 90 cm is equivalent to connecting 2 or 3 lengths in series and if the resistance of the 30 cm length is denoted by R Ω, then

$$\text{Resistance of 2 lengths (60 cm)} = R + R = 2\,R\,\Omega$$

$$\text{Resistance of 3 lengths (90 cm)} = R + R + R = 3R\,\Omega$$

from which it is evident that the resistance of the wire varies directly as its length, written

$$R \propto \ell.$$

If, however, two or more wires are connected in ***parallel*** (imagine the wires to be placed parallel to each other with their adjacent ends all connected together) and provided that the p.d. across the ends remains unchanged, it is clear that compared with one wire, two will carry twice the current therefore the total resistance must be halved, for by Ohm's Law since $V = IR$ and V is constant, then if I doubles to 2I and R_s and R_d denote the single and double wire resistances,

$$IR_s = 2IR_d \quad \text{(divide both sides by I)}$$

$$R_s = 2R_d \quad \text{or} \quad R_d = \tfrac{1}{2}R_s$$

Similarly three wires carry three times the current and the net resistance is reduced to one-third.

Now what has been done in effect is to increase the cross-sectional area of the wire by 2 or 3 times, thus the resistance varies inversely(A4.2.2) as the area of cross-section (a) and we can write

$$R \propto \frac{1}{a}$$

and the two relationships can be combined as

$$R \propto \frac{\ell}{a}$$

where R = resistance of wire or material, ℓ = the length, a = the area of cross-section which need not necessarily be circular as in the case of most wires.

However, for a wire or other conductor of circular cross-section, radius r,

$$a = \pi r^2 \text{ (A5.1)}$$

and in terms of the wire diameter (d), since d = 2r, i.e. r = d/2

$$a = \pi\left(\frac{d}{2}\right)^2 = \frac{\pi d^2}{4} \quad \text{(A4.1.2)}$$

and since $\pi/4$ is a constant,

$$a \propto d^2$$

from which, $R \propto \dfrac{\ell}{d^2}$

Summarizing: the resistance of a wire or material varies directly with its length and inversely with the area of cross-section, i.e. $R \propto \ell/a$. For circular cross-section wires, etc. $R \propto \ell/d^2$.

EXAMPLE:
Copper wire tables quote the resistance of 2 mm diameter copper wire as 0.005488 Ω per metre. What is the resistance of a length of 30 cm?

Since R varies directly with ℓ,

If resistance of 100 cm = 0.005488 Ω (= 0.005488 x 10^6 μΩ = 5488 μΩ), then resistance of

$$1 \text{ cm} = \frac{5488}{100} = 54.88\ \mu\Omega$$

and resistance of

$$30 \text{ cm} = 54.88 \times 30 = \underline{1646.4\ \mu\Omega}$$

or using a mathematical sequence(A4.2.2)

$R_1 = 0.005488\ \Omega$ $\qquad\ell_1 = 100$ cm

$R_2 = ?$ $\qquad\ell_2 = 30$ cm

Then

$$\frac{R_1}{R_2} = \frac{\ell_1}{\ell_2}$$

$$\therefore \frac{0.005488}{R_2} = \frac{100}{30} \text{ (cross multiply)}$$

$$100\, R_2 = 30 \times 0.005488$$

$$\therefore R_2 = \frac{30 \times 0.005488}{100} = 0.0016464\ \Omega = \underline{1646.4\ \mu\Omega}$$

EXAMPLE:
If 2 mm diameter copper wire has a resistance of 0.005488 Ω/metre, what will be the resistance of 0.2 mm diameter wire?

$R_1 = 0.005488\ \Omega \qquad d_1 = 2\text{ mm}$

$R_2 = ? \qquad d_2 = 0.2\text{ mm}$

Then

$$\frac{R_1}{R_2} = \frac{{d_2}^2}{{d_1}^2}$$

(remember the resistance is *inversely* proportional to d^2).

$$\therefore \frac{0.005488}{R_2} = \left(\frac{d_2}{d_1}\right)^2 = \left(\frac{0.2}{2}\right)^2 = 0.1^2 = 0.01$$

$$\therefore 0.01\ R_2 = 0.005488$$

$$\therefore R_2 = \frac{0.005488}{0.01} = \underline{0.5488\ \Omega}$$

EXAMPLE:
The resistance of a length of circular cross-section wire is 200 Ω. What is the resistance of half this length, having twice the diameter?

Let first length = ℓ, first diameter = d, first resistance = R_1, then second length = ℓ/2, second diameter = 2d, second resistance = R_2, and

$$\frac{R_1}{R_2} = \frac{\dfrac{\ell}{d^2}}{\dfrac{\ell}{2 \times 4d^2}} = \frac{\ell}{d^2} \times \frac{8d^2}{\ell} = 8$$

$$\therefore \frac{200}{R_2} = 8$$

$$\therefore R_2 = \frac{200}{8} = \underline{25\ \Omega}$$

3.4.2 Variation with material

That the resistance is different for each element or compound is clear from Section 1.2.2. Ohm's Law gives us a method of measuring resistance for if a current is made to flow through the material and the p.d. across it measured, the value of the resistance can be calculated. One of the traps mentioned earlier is that the resistance itself may depend on the magnitude of the measuring current since the resistance of some materials is current sensitive while most undergo a resistance change due to the heating effect of the current. Evidently the measuring current must be small, and with modern measuring equipment this is so.

For uniformity, because, as shown in the preceding section, resistance varies with dimension, an internationally agreed shape is used as a standard. This shape is a metre cube and the resistance is that measured between any two opposite faces, for example, between the wires A and B in Fig. 3.8. Obviously a block of material this size is not cut and measured, imagine the cost and weight of a metre cube of copper!

EXAMPLE:

It was suggested in Section 1.2 that there could be as many as 10^{23} free electrons in a piece of copper wire 30 cm long and 2 mm diameter. For practice with exponents, volumes and areas let us calculate how many free electrons there might be in a metre cube of copper.

Volume of wire = area of cross-section × length[(A5.3)]
(diameter = 2 mm = 0.2 cm)

$$= \left(\pi \times \frac{0.2^2}{4}\right) \times 30 \text{ cubic cm } (cm^3)$$

$$= 0.9425 \text{ cm}^3 = 0.9425 \times 10^{-6} \text{ cubic metres } (m^3)$$

(Note the apparent change in terms here – we have defined resistivity in terms of a metre cube but calculations of volumes are in cubic metres. A metre cube has a volume of one cubic metre but is of the particular shape used to define resistivity which depends on both shape and volume. A volume of one cubic metre can have any shape.)

Then, approximate number of free electrons in a metre cube (i.e. cubic metre) of copper:

$$= \frac{1}{0.9425 \times 10^{-6}} \times 10^{23} = \frac{10^6}{0.9425} \times 10^{23}$$

$$= \frac{1}{0.9425} \times 10^{29} \approx \underline{1.06 \times 10^{29}}$$

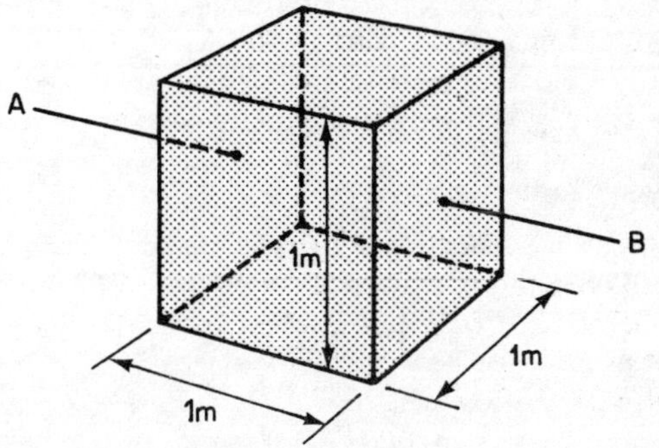

Fig. 3.8 Metre cube.

Any regular shape can be used for measurement and the resistance of the standard cube calculated as in the section above. The unit derived from this standard measurement is called an *ohm-metre* (Ωm) and with materials used for conductors the actual resistance value is very small, hence a more convenient unit, the microhm-metre (μΩm) may be used, there being 10^6 μΩm to 1 Ωm. These are measurements of the ***resistivity*** of the material.

For a little more practice in handling large numbers let us assume that the resistance of a 1 metre length of copper wire of diameter 5 mm has been measured and that the result is 0.0008781 Ω. We wish to calculate the resistivity in Ωm or μΩm.

Let R_1 be the measured result, ℓ_1 and a_1 the length and cross-sectional area of the sample wire, R_2 be the resistivity, ℓ_2 and a_2 the length and cross-sectional area of the standard cube.

Then

$$\ell_1 = 1\text{ m}, \quad a_1 = \frac{\pi {d_1}^2}{4} = \frac{\pi \times 5^2}{4} = 19.635\text{ sq.mm } (\text{mm}^2)$$

$$= 19.635 \times 10^{-6}\text{ sq. m } (\text{m}^2)$$

$$\ell_2 = 1\text{ m}, \quad a_2 = 1\text{ m}^2$$

$$\frac{R_1}{R_2} = \frac{\ell_1 a_2}{\ell_2 a_1}$$

$$\therefore \quad \frac{0.0008781}{R_2} = \frac{1 \times 1}{1 \times 19.635 \times 10^{-6}} = \frac{10^6}{19.635}$$

$$\therefore \quad R_2 = \frac{19.635 \times 0.0008781}{10^6}\ \Omega\text{m}$$

$= 19.635 \times 0.0008781\ \mu\Omega m = \underline{0.01724\ \mu\Omega m}$

i,e, the resistance between points A and B (Fig. 3.8) when the material is copper is 0.01724 $\mu\Omega$.

By expressing resistivity in this standard way, we have a ready comparison between all materials used in electronics, for example, very approximately:

Conductors:		
	Silver, Copper	0.01 – 0.02 $\mu\Omega m$
	Tin, Lead	0.1 – 0.2 "
	Mercury	about 1.0 "
Semiconductors:		0.01 – 100 Ωm
Insulators:		
	Ceramics, Polythene	10^{15} "
	Mica	$10^{14} - 10^{15}$ "
	Glass	10^{11} "
	Dry Paper	$10^{9} - 10^{10}$ "

The figures also enable us to calculate the resistance of any other shape, (e.g. a wire) and if we denote the resistivity by the Greek letter ρ (rho, pronounced as in rope), then

$$R = \rho\frac{\ell}{a}$$

where R = resistance in ohms, ℓ = length in metres, a = area of cross section in square metres and ρ = resistivity in Ωm.

This is an important formula to remember for practical resistances can be calculated from tables of resistivities.

EXAMPLE:
What is the resistance of an aluminium rod 80 cm long and 0.5 cm square cross-section if the resistivity of aluminium is 0.028 $\mu\Omega m$.

$$R = ? \qquad \rho = 0.028 \times 10^{-6}\ \Omega m$$

$$\ell = 80 \times 10^{-2}\ m$$

$$a = (0.5 \times 10^{-2})^2 = 0.25 \times 10^{-4}\ m^2$$

Then:

$$R = \rho\frac{\ell}{a} = \frac{0.028 \times 10^{-6} \times 80 \times 10^{-2}}{0.25 \times 10^{-4}}$$

$$= \frac{0.028 \times 80}{0.25} \times \frac{10^{-6} \times 10^{-2}}{10^{-4}}$$

$$= 8.96 \times 10^{-4}\ \Omega$$

$$= 8.96 \times 10^{-4} \times 10^{6}\ \mu\Omega = \underline{896\ \mu\Omega}$$

(a reminder – ρ must be in ohm-metres, ℓ in metres, a in square metres, the value calculated for R follows in ohms).

EXAMPLE:
Calculate the resistance per metre of fine silver wire, 0.1 mm diameter if the resistivity of silver is 0.016 $\mu\Omega$m.

$$R = ? \qquad \rho = 0.016 \times 10^{-6}\ \Omega m \qquad \ell = 1\ m$$

$$a = \pi \times \frac{0.1^2}{4} = \frac{\pi \times 0.01}{4} = \frac{0.031416}{4}$$

$$= 0.007854\ mm^2 = 0.007854 \times 10^{-6}\ m^2$$

Then

$$R = \rho\frac{\ell}{a} = 0.016 \times 10^{-6} \times \frac{1}{0.007854 \times 10^{-6}}$$

$$= \frac{0.016}{0.007854} = \underline{2.037\ \Omega}$$

3.4.3 Variation with temperature

For most conductors the resistance rises with temperature. Section 1.2.3 explains this effect reasonably satisfactorily by showing that current flow generates heat through electrons giving up energy (gained from the e.m.f.) in their collisions with atoms.

As each electron loses energy, its speed falls and it is therefore no longer as efficient a charge carrier, the e.m.f. thus has greater difficulty in causing the current to flow, that is, the resistance of the conductor has increased.

In those materials, mainly insulators, which exhibit a decrease in resistance as temperature rises, conditions are somewhat different. Although very few electrons can take part in the conduction process, more are freed as temperature rises thus resistance tends to fall. The effect of energy transfer as for conductors is of less importance because of the very few collisions occurring.

For practical work a *Temperature Coefficient of Resistance* is used for each different material and this is defined as the fractional change in resistance per unit change in temperature. Most conveniently the temperature scale used is Centigrade (C) on which melting ice is at 0° and boiling water 100°. The temperature coefficient of resistance (denoted by σ, the Greek letter sigma) is quoted at 0°C and the resistance changes this fraction of its value at 0°C, for every degree change in temperature. As an example, using the simple figure of $\sigma = 0.1$ for a particular material, if the resistance at 0°C is 1 Ω, then a fractional change of 0.1 brings the resistance to 1.1 Ω at 1°C, 1.2 Ω at 2°C, etc., the increase in resistance being calculated by adding the fractional change multiplied by the number of degrees temperature change, expressed mathematically as

$$R_t = R_0(1 + \sigma t) \qquad \text{(i)}$$

where R_t is the resistance at t°C, R_0 is the resistance at 0°C, σ is the temperature coefficient of resistance, and t is the temperature at which the resistance is required.

Referring again to the above example, for $\sigma = 0.1$, when t = 1°C, the resistance of 1 Ω has risen to $1[1 + (0.1 \times 1)] = 1.1\ \Omega$, and when t = 2°C the resistance of 1 Ω has risen to $1[1 + (0.1 \times 2)] = 1.2\ \Omega$ as shown.

For materials for which the resistance decreases as temperature rises, σ is negative and thus $\sigma = -0.1$ would give when t = 1°C, the resistance of 1 Ω has fallen to $1[1 + (-0.1 \times 1)] = 0.9\ \Omega$, and when t = 2°C the resistance of 1 Ω has fallen to $1[1 + (-0.1 \times 2)] = 0.8\ \Omega$, showing that the formula covers all cases. It is generally of suffiicient accuracy up to about 100°C; for temperatures outside the range 0 – 100°C accurate calculations are possible but the temperature coefficient becomes more complicated.

If the resistance (R_1) of a material is known at some temperature other than 0°C (say t_1) and it is required to calculate the resistance (R_2) at a second temperature (t_2) it would appear that the formula has to be used twice, the first time to obtain the resistance (R_0) at 0° C, the second to calculate the resistance at temperature t_2. To avoid the double calculation the following formula gives a reasonably accurate answer:

$$R_2 = R_1[1 + \sigma(t_2 - t_1)] \qquad \text{(ii)}$$

In the following example both methods are used.

EXAMPLE:
The resistance of a length of copper wire measures 70 Ω at a temperature of 10°C. What will the resistance be at 30°C? The temperature coefficient of resistance of copper can be taken as +0.004/°C.

$R_0 = ?$ $\quad R_1 = 70\ \Omega$ $\quad t_1 = 10°C$

$R_2 = ?$ $\quad t_2 = 30°C$ $\quad \sigma = +0.004/°C$

Using the first formula

$$R_1 = R_0(1 + \sigma t_1)$$

$$\therefore \quad 70 = R_0[1 + (0.004 \times 10)] = R_0(1 + 0.04)$$

$$\therefore \quad 70 = R_0 \times 1.04$$

$$R_0 = \frac{70}{1.04} = 67.3077\ \Omega$$

Also $R_2 = R_0(1 + \sigma t_2) = 67.3077[1 + (0.004 \times 30)]$

$$= 67.3077(1 + 0.12) = 67.3077 \times 1.12$$

$$= \underline{75.3846\ \Omega}$$

Using the second formula:

$$R_2 = R_1[1 + \sigma(t_2 - t_1)] = 70[1 + \sigma(30 - 10)]$$

$$= 70(1 + 0.08) = \underline{75.6\ \Omega}$$

the greater ease but slight inaccuracy of the second formula is therefore apparent.

EXAMPLE:
If the temperature coefficient of resistance of nickel is taken as $6 \times 10^{-3}/°C$ and the resistance of a length of nickel wire measures 275.59 Ω at 0°C, at what temperature does the resistance rise to 350 Ω?

$R_0 = 275.59\ \Omega$ $\quad R_t = 350\ \Omega$ $\quad t = ?$

$\sigma = 6 \times 10^{-3}$ (i.e. 0.006)

Then

$$R_t = R_0(1 + \sigma t)$$

$$\therefore \quad 350 = 275.59(1 + 0.006t)$$

$$\therefore \quad \frac{350}{275.59} = 1 + 0.006t$$

$$\therefore \quad 1.27 = 1 + 0.006t$$

$$0.27 = 0.006t$$

$$\therefore \quad \frac{0.27}{0.006} = t$$

i.e., $t = 45$, hence the resistance rises to 350 Ω at 45°C

EXAMPLE:
A carbon rod has a resistance of 2000 Ω at room temperature. Approximately what will its resistance be at 62°C if the temperature coefficient of resistance is –0.0005/°C and room temperature is taken as 20°C?

Asking for an approximate answer allows us to use formula (ii)

$R_1 = 2000\ \Omega \qquad t_1 = 20°C$

$R_2 = ? \qquad t_2 = 62°C \qquad \sigma = -0.0005/°C$

Then:

$$R_2 = R_1[1 + \sigma(t_2 - t_1)]$$

$$= 2000[1 - 0.0005(62 - 20)]$$ (remember that σ is negative)

$$= 2000[1 - (0.0005 \times 42)]$$

$$= 2000(1 - 0.021) = 2000 \times 0.979 = \underline{1958\ \Omega}$$

This example demonstrates the fall in resistance of certain materials as temperature rises. Special design is sometimes required in electronic circuits when such materials are used to avoid their destruction. Current flow produces heat, causing a temperature rise, whereupon the resistance falls and the current accordingly increases. This creates even more heat with a further lowering of resistance and so on until with certain devices, especially transistors, the component is destroyed by overheating.

σ for most conductors lies within the range 0.003 – 0.005/°C but when a stable resistance value is required, values of σ as low as 0.000014 (14×10^{-6})/°C can be produced as in, for example, Eureka, Constantan or Manganin which are alloys from which resistance wires of high stability are made.

3.4.4 Practical resistors

It is obvious that all electronic circuits contain resistance, for example, as the internal resistance of a battery or even of connecting wires (there is one exception only, resistance does disappear at extremely low temperatures, but so low that we shall not be concerned). Such resistances can be measured or calculated and if unwanted, may be reduced to a certain degree but some will remain and frequently the circuit design has to make due allowance. However, in electronic circuit work both voltages and currents need to be controlled to bring about the desired effects and this is frequently arranged by the inclusion of *resistors* which are simply small devices having a certain value of resistance. To fully appreciate all aspects of rèsistors it is instructive to go through a simple design process not only to demonstrate what this section is about but also to give more practice with Ohm's Law and a newly introduced concept of *tolerances.*

Designing a resistor chain:

Transistors need a *bias* voltage for correct operation, meaning that the *base* of the transistor is maintained at a steady voltage relative to the *emitter.* These electrodes are shown in Fig. 3.9

and in this example it is assumed that the base must be held to −0.3 ± 30% V relative to the emitter. The biassing requirements are obtained by the use of two resistors R_1 and R_2 as shown. We will assume that the 6-volt battery has no internal resistance, that is, whatever current it provides, the terminal voltage remains at 6. The transistor emitter is connected to the battery + line which for convenience we label 0 and the top battery line is therefore −6 V relative to this. To obtain −0.3 V at the base we could use for R_1 a resistor of 19,000 Ω and for R_2, 1000 Ω. Let us check this first.

We know from Section 3.3.1 that the total value (R_T) of R_1 + R_2 is 19,000 + 1000 = 20,000 Ω and therefore by Ohm's Law, the current (I) through them is V/R_T, i.e. 6/20,000 amperes, and again, from Ohm's Law, the p.d. across R_2

$$= IR_2 = \frac{6}{20{,}000} \times 1000 = 0.3 \text{ V}$$

and clearly from Fig. 3.9 the voltage is negative relative to the 0 (+) line.

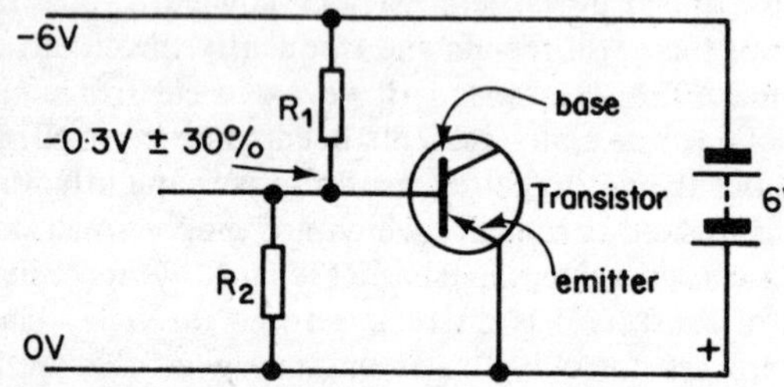

Fig. 3.9 Voltage divider chain.

Thus we demonstrate a use of two resistors to provide a transistor bias voltage (the design is in fact slightly more complicated because account may need to be taken of a tiny current flowing in the transistor base circuit).

Division of voltage in this way (by means of 2 resistors) is frequently required in design and, returning to the calculation above, if:

$$I = \frac{V}{R_T}, \quad \text{i.e. } I = \frac{V}{R_1 + R_2}$$

and the p.d. across $R_2 = IR_2 = \dfrac{VR_2}{R_1 + R_2}$

we notice that the current I has disappeared from the final expression and thus need not be calculated, so giving a general formula for the division of voltage in a 2-resistor series chain:

$$\text{Voltage across } R_1 = \frac{VR_1}{R_1 + R_2}$$

$$\text{Voltage across } R_2 = \frac{VR_2}{R_1 + R_2}$$

Continuing the example, the next step is in choosing the type of resistor to be used. There are many different types available but of these, for this particular purpose, the commonly used *composition* resistor is suitable, it is inexpensive, obtainable in a wide range of values, is small and easily soldered into the practical circuit. The resistance element is a mixture of carbon contained in a small tube with the two wire connexions, each a few cms long, embedded in the ends as shown in Fig. 3.10.

But, having chosen the type of resistor, other details must be specified:

(i) the actual value to be used – clearly no manufacturer could be expected to produce resistors having all values from, say, 1 ohm to 10 MΩ, a more limited number is quite feasible;

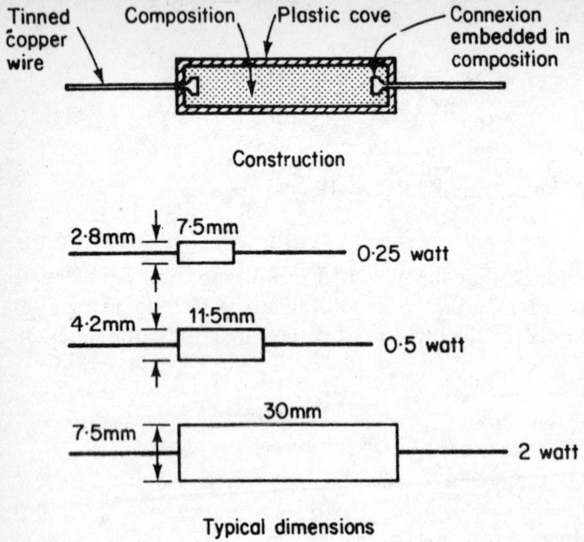

Fig. 3.10 Carbon composition resistors.

(ii) the "tolerance" of value – accuracy and cheapness do not come together so we must choose between the two;

(iii) the "power rating" – current will flow through the resistor and generate heat, too much may be destructive.

These are now considered in more detail.

Value:
In most cases a *preferred value*(A10.2) is suitable and from Table A10.1, R_2 is suited exactly because 1 kΩ is a preferred value, but for R_1 the nearest preferred value to 19 kΩ is 18 kΩ (10% only) or 18 kΩ or 20 kΩ (5%). But first we must determine the design voltage range, i.e.

$$0.3\text{ V} \pm 30\% = 0.3 + \left(\frac{30}{100} \times 0.3\right) \text{ to}$$

$$0.3 - \left(\frac{30}{100} \times 0.3\right)$$

i.e. 0.39 – 0.21 V ,

and having done this, consider the effect of both resistors R_1 and R_2 being at one extreme of the tolerance range. Suppose we first choose the most inexpensive combination which would be for R_1, 18 kΩ 10% and for R_2, 1 kΩ 20%. Then the actual value for R_1 could be between 18 + (0.1 x 18) and 18 – (0.1 x 18) kΩ, i.e. 19.8 kΩ and 16.2 kΩ, and for R_2, 1.2 kΩ and 0.8 kΩ.

If R_1 is high and R_2 low, the bias voltage will be at its lowest, and using the formula,

$$\text{Voltage across } R_2 = \frac{VR_2}{R_1 + R_2} \qquad V = 6\text{ V}$$

$$R_1 = 19.8\text{ k}\Omega$$

$$R_2 = 0.8\text{ k}\Omega$$

$$\text{Voltage across } R_2 = \frac{6 \times 0.8}{19.8 + 0.8}$$

(since all resistor values are in kilohms, we need not convert to ohms)

$$= 0.233\text{ V}$$

which is within the limits. Similarly, for the highest bias voltage:

$$\text{Voltage across } R_2 = \frac{6 \times 1.2}{16.2 + 1.2} = 0.414\text{ V}$$

which is *not* within the limits.

We might next try using a 10% tolerance resistor for R_2 where maximum tolerances produce a resistance range from 0.9 to 1.1 kΩ and again calculating the highest bias voltage using 1.1 kΩ for R_2 gives 0.382 V, which is within the design range.

Thus two resistors which produce –0.3 V ± 30% bias voltage from a 6 V supply could be

$$R_1 = 18\ \text{k}\Omega\ 10\%, \qquad R_2 = 1\ \text{k}\Omega\ 10\%$$

Remember that we have calculated these values so that if the practical resistors used in the circuit happened to be at the extremes of their tolerances (which is unlikely), the design requirements would still be satisfied. There is therefore no reason to use more expensive resistors of closer tolerances.

Power rating:
Having chosen the pair of resistors with their tolerances, there is one further step which is to calculate the power dissipated in each and hence be able to choose between the physical sizes offered. The commonly used sizes are 0.25 watt and 0.5 watt (typical dimensions are shown in Fig. 3.10), subminiature 0.125 and 0.1 watt are also obtainable as are larger, for example 1 or 2 watt. The calculation follows that shown in Section 3.1.1, great accuracy is seldom required especially for transistor circuit calculations where currents may be extremely small – let us see in this case.

There is a voltage V = 6 V across a total resistance of 19 kΩ, resulting in a current

$$I = \frac{6}{19{,}000}\ \text{A}$$

and since the voltage across R_1 is (6 – 0.3)V = 5.7 V, the power dissipated in R_1 = 5.7 x (6/19000) = 0.0018 watts, which is easily handled by the smallest practical resistor, in fact there is so much latitude that there is no point in further consideration of R_1 being at an extreme of tolerance.

Also power dissipated in $R_2 = 0.3 \times (6/19000)$, which must be even less.

Of course, we could have obtained 0.3 V bias by using much lower resistances, say $R_1 = 18\ \Omega$ and $R_2 = 1\ \Omega$. The power dissipated in R_1 would then be 1.8 watts for which a 2-watt resistor would be required. The unnecessary size and cost of R_1 alone would be a mark of poor design, but in addition there is a wasteful drain on the battery supply – however, these considerations are proper to transistor circuit design rather than as an illustration of the choice of practical resistors.

Marking codes:
A general coding system using colours to indicate value and tolerance is used for small resistors. Because components other than resistors may use the same code, the system is described in Appendix 10.

Resistor types:
Mention has been made of the carbon composition type which is perfectly adequate for a whole range of circuits. The advantages are low cost, small size and availability in a wide range of resistance values. There are however several disadvantages, mainly in stability (that is, capability of maintaining characteristics) which is poor both with time and temperature and in availability of power ratings, these are limited to about 2 watts. Thus other types of resistor are available, usually costing more but with better stabilities and accuracies. They are usually manufactured in the same tubular form and the resistance element may consist of carbon, metal film formed on a tube or of metal oxide.

Alternatively for the lower range of resistances, wire-wound types are available, using high resistivity wires (around 0.2 to 0.5 $\mu\Omega$m – Section 3.4.2) with a low temperature coefficient of resistance (Section 3.4.3) to minimize resistance change with increase of temperature. The wire may be coiled or spooled and left open or coated with vitreous (glasslike) enamel.

Variable resistors are usually made by using a *slider* which is a moving arm making contact with either different turns of a wire coil or over a carbon composition track.

Finally, resistors are also produced automatically within complete but encased circuits, such as the modern *integrated circuit* in which all the components and wiring are put together to make up the complete circuit on an unbelievably small *chip* of metal. Needless to say, resistors in these circuits are inaccessible.

3.4.5 Resistances in parallel

One of the two simple modes of interconnecting resistances, that is, in series, has been discussed in Section 3.3.1 showing that the values of the individual resistances add up to produce

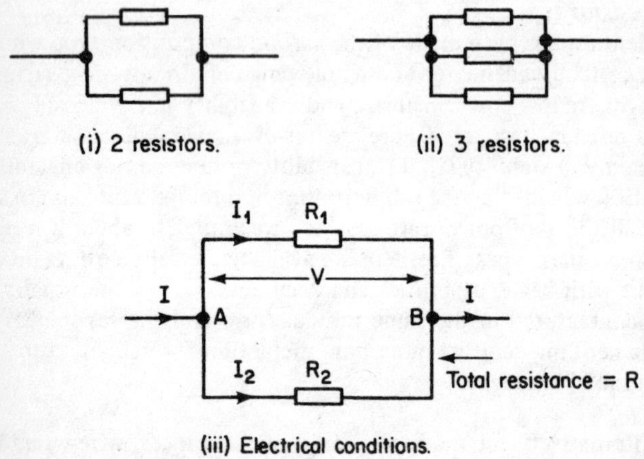

Fig. 3.11 Resistances in parallel.

the value of the total. Resistances connected in parallel that is, with adjacent ends connected together as shown in Fig. 3.11, have a slightly more complicated formula for the total value.

Consider two resistors as in Fig. 3.11(i) and let the electrical conditions be as shown in (iii), with a current of I amperes flowing (direction does not matter). At point A the current I will divide in some manner and if I_1 flows through resistor R_1 and I_2 through R_2, evidently $I = I_1 + I_2$. I_1 and I_2 combine at point B and continue together as the main current I.

Now, across points A and B a voltage must be developed because a current is flowing through resistance, this is labelled V and it is developed across both R_1 and R_2, hence from Ohm's Law

$$I_1 = \frac{V}{R_1} \quad \text{and} \quad I_2 = \frac{V}{R_2}$$

thus if one resistance is lower than the other, it will carry a greater current. Now if R is the total resistance of R_1 and R_2 in parallel,

$$I = \frac{V}{R}$$

and since $I = I_1 + I_2$,

$$\frac{V}{R} = \frac{V}{R_1} + \frac{V}{R_2}$$

we can divide both sides by V, giving

$$\frac{1}{R} = \frac{1}{R_1} + \frac{1}{R_2}$$

and similarly for 3 or more resistances in parallel

$$\frac{1}{R} = \frac{1}{R_1} + \frac{1}{R_2} + \frac{1}{R_3} + \ldots$$

that is, the reciprocal of the total resistance is equal to the sum of the reciprocals of the individual resistances.

For the 2-resistor combination, a little manipulation of the formula enables us to obtain the total resistance more directly for since

$$\frac{1}{R} = \frac{1}{R_1} + \frac{1}{R_2}$$

adding the two fractions on the r.h.s. together with a common denominator $R_1 R_2$

$$\frac{1}{R} = \frac{R_2 + R_1}{R_1 R_2}$$

$$\therefore \quad R = \frac{R_1 R_2}{R_1 + R_2}$$

EXAMPLE:
Two preferred-value resistors of 68 Ω and 33 Ω are connected in parallel. What is the resultant resistance?

$R_1 = 68\ \Omega \qquad R_2 = 33\ \Omega \qquad R = ?$

$$R = \frac{R_1 R_2}{R_1 + R_2} = \frac{68 \times 33}{68 + 33} = \frac{2244}{101} = \underline{22.2\ \Omega}$$

EXAMPLE:
What value resistor should be connected in parallel with a 47 kΩ resistor to produce a total resistance of 39 kΩ?

$$R = 39\text{ k}\Omega \qquad R_1 = 47\text{ k}\Omega \qquad R_2 = ?$$

Working directly in kΩ,

$$\frac{1}{39} = \frac{1}{47} + \frac{1}{R_2}$$

$$\therefore \quad \frac{1}{R_2} = \frac{1}{39} - \frac{1}{47} = \frac{47 - 39}{39 \times 47} = \frac{8}{39 \times 47}$$

$$\therefore \quad 8R_2 = 39 \times 47$$

$$R_2 = \frac{39 \times 47}{8} = 229.13\text{ k}\Omega\text{, i.e. } \underline{\text{approx. } 230\text{ k}\Omega}$$

EXAMPLE:
Three resistors of 15, 22 and 33 ohms are connected in parallel. What is the joint resistance?

$$R_1 = 15\ \Omega \qquad R_2 = 22\ \Omega \qquad R_3 = 33\ \Omega$$

$$R = ?$$

Then:

$$\frac{1}{R} = \frac{1}{15} + \frac{1}{22} + \frac{1}{33} = \frac{22 + 15 + 10}{2 \times 3 \times 5 \times 11} = \frac{47}{330}$$

$$\therefore \quad R = \frac{330}{47} = \underline{7.02\ \Omega}$$

Factorization was moderately easy in this case but generally for this type of calculation when neither calculator nor slide-rule is available, but a general book of tables is, the table of reciprocals is useful:

Reciprocal of 15	=	0.06667
" " 22	=	0.04545
" " 23	=	0.03030
Total reciprocal	=	0.14242,

this equals 1/R, therefore, reversing the procedure in the reciprocal tables gives R = 7.02 Ω .

EXAMPLE:
A 12 V car battery is connected to two headlamps labelled 36 watts and 24 watts. Find the total resistance, the total current and the current through each lamp.

The circuit diagram is as shown in Fig. 3.12.

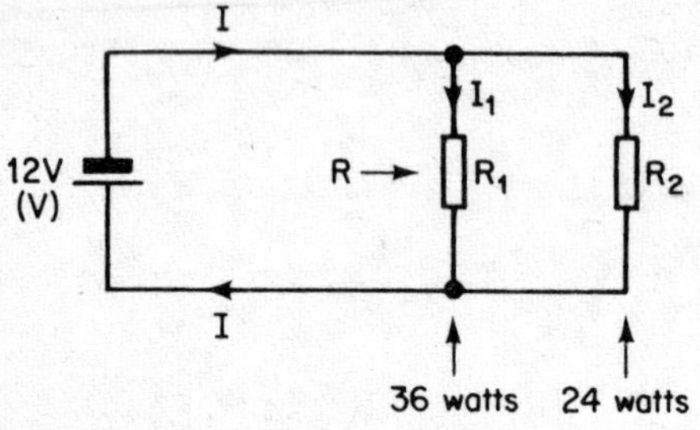

Fig. 3.12 Car battery connected to lamps.

Since:

$$\text{Power} = \text{Voltage} \times \text{Current}$$

$$36 = 12 \times I_1 \qquad\qquad 24 = 12 \times I_2$$

$\therefore \quad \underline{I_1 = 3\text{ A}} \quad \text{and} \quad \underline{I_2 = 2\text{ A}}$

and since

$$\text{Resistance} = \frac{\text{Voltage}}{\text{Current}}$$

$$R_1 = \frac{12}{3} = 4\,\Omega \qquad R_2 = \frac{12}{2} = 6\,\Omega$$

$$\therefore \quad \text{Total resistance, } R = \frac{4 \times 6}{4 + 6} = \underline{2.4\,\Omega}$$

$$\therefore \quad \text{Total current} = \frac{12}{2.4} = \underline{5\text{ A}}$$

which could also be obtained from $I_1 + I_2$, i.e. $3 + 2 = \underline{5\text{ A}}$.

This example leads to a general formula for the division of current in two parallel resistors. Referring to Fig. 3.12, to find I_1:

From Ohm's Law

$$V = IR \quad \text{and as previously shown,} \quad R = \frac{R_1 R_2}{R_1 + R_2}$$

$$\therefore \quad V = I \times \frac{R_1 R_2}{R_1 + R_2}$$

$$\text{Also } I_1 = \frac{V}{R_1}$$

$$\therefore \quad I_1 = I \times \frac{R_1 R_2}{R_1 + R_2} \times \frac{1}{R_1} \quad (\text{cancel } R_1)$$

$$\therefore \quad I_1 = I \cdot \frac{R_2}{R_1 + R_2}$$

Similarly

$$I_2 = I \cdot \frac{R_1}{R_1 + R_2}$$

that is, the current in each path is equal to the main current multiplied by the resistance of the *opposite* path and divided by the sum of the two path resistances.

When n resistors of the same value are connected in parallel, the total value is equal to the single resistor value divided by n, e.g. for

$$R_1 = 20\,\Omega \qquad R_2 = 20\,\Omega \qquad R_3 = 20\,\Omega$$

$$R_4 = 20\,\Omega \qquad n = 4 \qquad R = ?$$

$$\frac{1}{R} = \frac{1}{20} + \frac{1}{20} + \frac{1}{20} + \frac{1}{20} = \frac{4}{20}$$

$$\therefore \quad R = \frac{20}{4}$$

which demonstrates the rule.

3.4.6 *Resistances in series–parallel*

Fig. 3.13(i) shows such an arrangement with two series resistances R_1 and R_4 in series with two parallel networks A and B.

Each parallel network is considered separately and reduced to its single series value, the total resistance R is then the sum of the series values as shown in (ii).

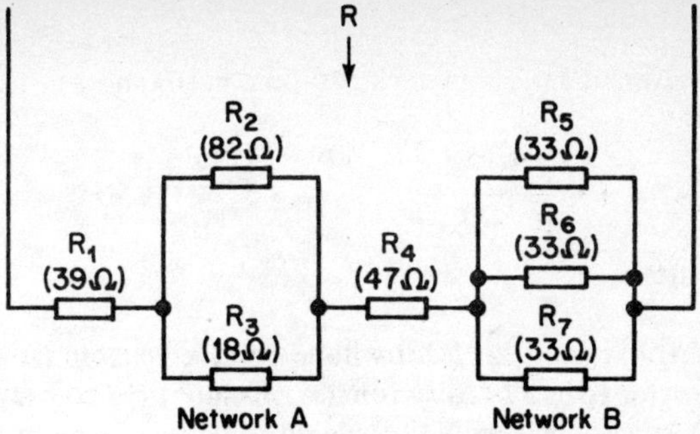

(i) Complete network.

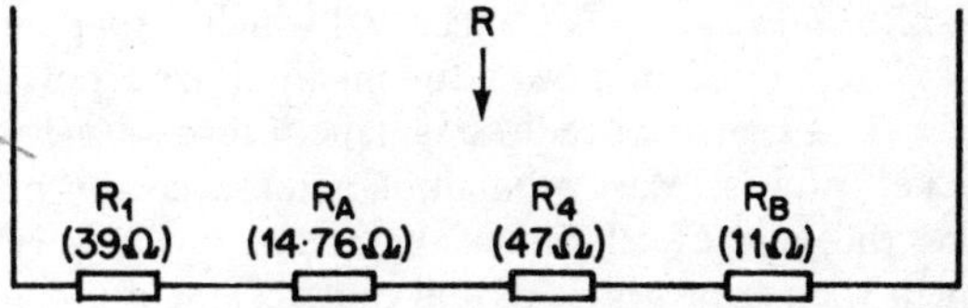

(ii) Equivalent series-only network.

Fig. 3.13 Series-parallel arrangement of resistances.

EXAMPLE:
If the resistances in Fig. 3.13(i) have the values shown in brackets, what is the total resistance, R, of the whole network?

Let the total resistance of Network A be R_A and of Network B be R_B

$$R_A = \frac{82 \times 18}{82 + 18} = 14.76\ \Omega$$

$$R_B = \frac{33}{3} = 11\,\Omega$$

The equivalent series network is shown in (ii) and

$$R = 39 + 14.76 + 47 + 11 = \underline{111.76\,\Omega}$$

3.5 Batteries

Referring back to Fig. 2.2 it will be observed that what many people refer to as a *battery* is in fact technically known as a *cell*. A battery is usually considered to be two or more cells connected together.

3.5.1 Primary and secondary cells

Fig. 2.2 also shows a dry Leclanché cell which is found everywhere as the source of power for medium-sized portable devices such as transistor radio sets, tape recorders, calculators and pocket torches. Mass production and low cost of materials makes this type of cell one of the cheapest. For very small equipment such as hearing-aids and cameras, the dry mercury cell is used, it has a voltage of 1.35 which is slightly lower than that of the Leclanché at 1.5.

These are the most commonly found *primary* cells, in which the chemical process which takes place in the electrolyte is not reversible, thus when exhausted the cell is discarded. The alternative type of cell is the *secondary* in which the chemical action is reversed by *recharging*. A common example of a secondary cell is the one used in car batteries, consisting essentially of plates made of a lead framework and containing lead compounds, immersed in an electrolyte of dilute sulphuric acid. Current is passed through the cell to *form* the plates, whereupon the active material changes and the cell is then able to function by providing a current. The cell voltage is about 2 and because it has a relatively low internal resistance a large current can be supplied without an excessive fall in terminal voltage. Because the electrolyte is a liquid, the cell is classified as a *wet* type. A "dry" type of rechargeable cell is the nickel-cadmium, it has a voltage of about 1.25 and is mainly used in portable equipment where recharging secondary cells is much less expensive than replacement of primary ones, for example in calculators, photographic flash equipment, small power tools and electric shavers.

3.5.2 Connexion of cells

Most frequently encountered is the series connexion as shown in Fig. 3.14(i) and (ii) where the battery voltage is the sum of the individual cell voltages. In this method adjacent cells are connected together by linking one terminal of each of opposite polarity. If n similar cells are connected in series, the battery voltage is n times the cell voltage. Similarly because the cell internal resistances are effectively in series and therefore additive, the battery internal resistance is n times the cell resistance.

EXAMPLE:

A 6 V battery consists of 4 1.5 V (e.m.f.) cells, each having an internal resistance of 2.5 Ω. What current will flow through a lamp having a resistance of 50 Ω? What will be the actual terminal voltage (p.d.) of each cell and of the battery when working?

Battery e.m.f. = 6 V

Battery internal resistance = 4 x 2.5 = 10 Ω

Total circuit resistance = 50 + 10 = 60 Ω

∴ From Ohm's Law

$$\text{Current} = \frac{6}{60}\ \text{A} = \underline{100\ \text{mA}} \quad \text{(i.e. current through lamp).}$$

Internal voltage drop of each cell

= Current x internal resistance = 0.1 A x 2.5 Ω

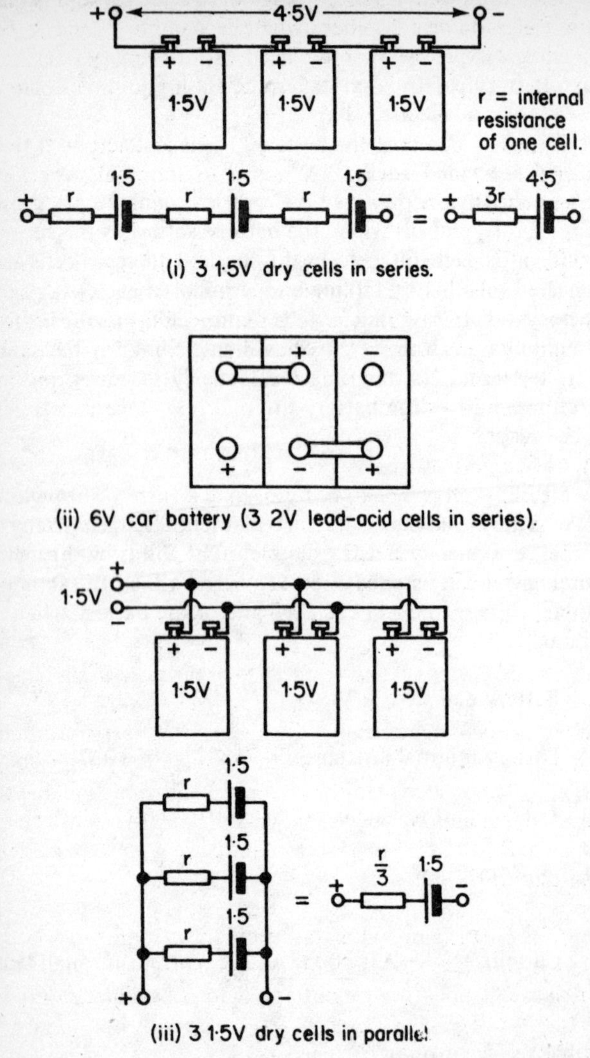

Fig. 3.14 Cells connected in series and parallel.

$= 0.25$ V

$\therefore$ Cell p.d. $= 1.5 - 0.25 = \underline{1.25 \text{ V}}$

and Battery p.d. $= 4 \times 1.25 = \underline{5.0 \text{ V}}$

i.e. of the 6 V e.m.f. of the battery, 1 V is lost across its own internal resistance leaving 5 V across its terminals when a current of 100 mA flows.

The parallel connexion is shown in Fig. 3.14(iii) which shows that for similar cells the battery and cell voltages are the same, however the internal resistance is reduced, that is, the battery can supply a heavier current than a single cell for the same fall in terminal p.d. The capacity of the battery is also n times the capacity of a single cell.

3.5.3 Charging secondary cells

A secondary cell is recharged by passing a current through it in the opposite direction to that of the discharge current. In Fig. 3.15 is shown, for example, a 12 V car battery discharging through a load, R (which might be the ignition coil, headlamps, starter, etc.).

The circuit current being from – to + through the load, must be flowing through the battery itself from + to –. Therefore to force a current through in the opposite direction, charges must be applied to the terminals of such value that they are greater than the charges already existing due to the chemical action, that is, the charging source must be connected with polarity as shown and have a voltage greater than that of the battery.

The following example illustrates this and also includes a calculation showing how the internal resistance of a current source such as a battery or battery charger, can be determined.

EXAMPLE:
In the charging circuit of Fig. 3.15, the voltage of the battery charger when measured with no load is 15 V. When measured

across a test load of 5 Ω, the voltage is 14.26 V. What will the charging current be when the charger is connected to the battery?

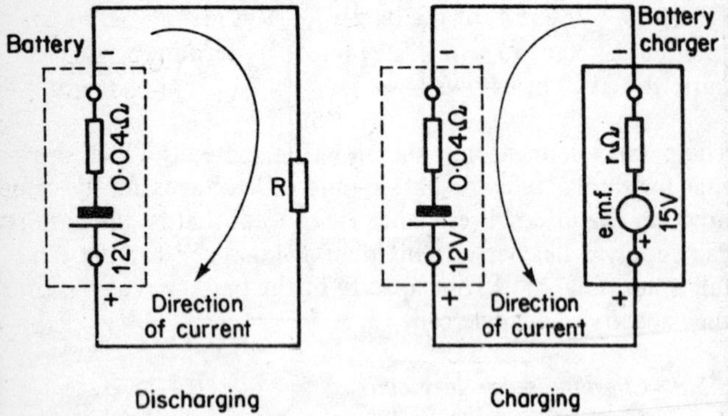

Fig. 3.15 12V car battery on discharge and charge.

Measuring the voltage of the charger off load gives the value of the e.m.f. (assuming that the measuring device takes no current – practical voltmeters take so little that it can be ignored), i.e. 15 V.

Let the internal resistance of the charger = rΩ

If 14.26 V is measured across a 5 Ω load, then circuit current

$$= \frac{14.26}{5} = 2.852 \text{ A}$$

and voltage drop across internal resistance, r

$$= 15 - 14.26 = 0.74 \text{ V}$$

(these conditions are illustrated in Fig. 3.16 which shows how the sum of the p.d.s in a circuit is equal to the e.m.f.)

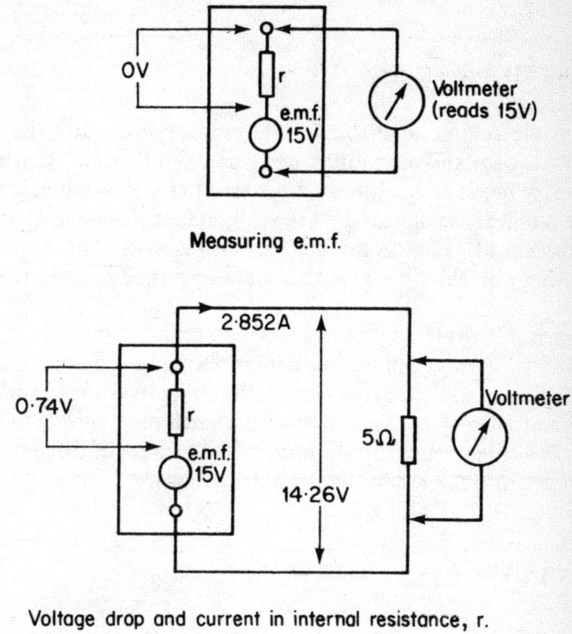

Fig. 3.16 Calculation of internal resistance of a current source.

$$\therefore \quad \text{Value of r} = \frac{0.74}{2.852} = 0.26\ \Omega$$

Then: Effective e.m.f. in the charging circuit of Fig. 3.15 (note that the e.m.f.s are in opposition)

$$= 15 - 12 = 3\ \text{V}$$

$$\text{Total circuit resistance} = 0.26 + 0.04 = 0.3\ \Omega$$

$$\therefore \text{Charging current} = \frac{3\text{ V}}{0.3\ \Omega} = \underline{10\text{ A}}$$

3.6 Electrical Safety

One electrical circuit should interest everybody, it is the one which contains the human body as one of the resistances. Safety needs to be discussed for electricity is well known to be a killer. Having studied the simple electric circuit it is now possible to examine the mechanism of shock and thereby perhaps be able to prevent accident to ourselves and others.

3.6.1 Earthing

The earth on which we live comprises many different materials, from solid rock to water, each having its own resistivity; this alone makes calculation of the earth resistance between two distant points almost impossible because of the unknown values of ρ for use in the formula (Section 3.4.2)

$$R = \rho \frac{\ell}{a}$$

The length (ℓ) of the path between two points may be measurable but a (the area of cross-section) is indeed complicated, embracing as it does, the whole earth, but we can certainly look upon it as being so enormous that even when ρ is high (say, for rock or dry sand), R will be quite low and earth resistance measurements prove this to be generally so, in fact of the order of a few ohms.

Early telegraph circuits dispensed with one wire by using the earth instead, at each telegraph station a connexion to earth was made by burying a metal plate, with the circuit as shown in Fig. 3.17.

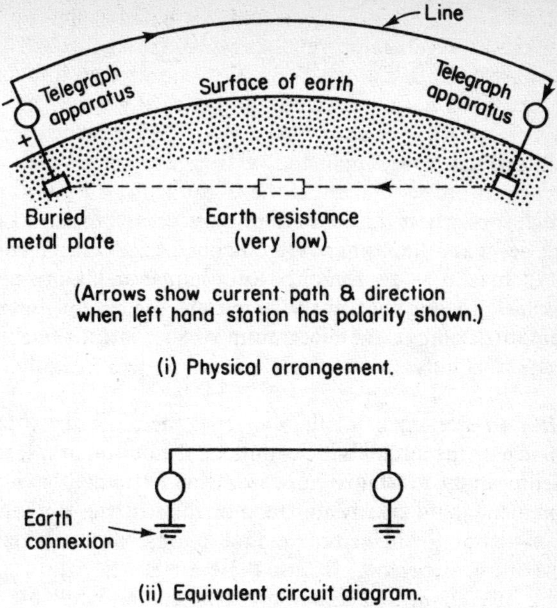

Fig. 3.17 Single wire telegraph circuit with earth return.

This is an example of using the earth as a low resistance conductor although there are disadvantages. Earthing may also be used with the power mains, not to carry circuit current but as a safety measure to prevent voltages higher than normal occurring at the consumer's premises. We cannot treat this in greater detail yet but this little knowledge is essential for appreciation of the safety aspects discussed below.

3.6.2 Electric shock

Recalling the fact that all matter is composed of atoms and electrons, it is evident that the human body has the capacity to carry an electric current, depending on its resistance, just as a wire does. The real danger from electricity arises from its interference with the heart beat, not through damage to the heart itself. The main function of a heart is to pump, but it is told when to do this by a *pacemaker* situated at the top. The pacemaker generates impulses which travel through the heart muscle fibres and when the fibres receive an impulse, the heart beats once. In electronic parlance we call this *triggering*, for it is not unlike the action of a gun which needs a small pressure to be applied to its trigger to release the much greater power of the bullet. If the trigger jams, the gun cannot be fired, similarly if the pacemaker pulse ceases, the heart no longer beats and the owner very quickly dies. An electronic current flowing across the chest of a human being of only about 100 mA (even less than the current needed by a small pocket-torch bulb) can be fatal simply by its interference with the triggering pulse.

The key to assessment of the danger is therefore the total resistance of the circuit which contains the person as a series element. Body resistance however, is an extremely variable factor, depending greatly on the moistness of the skin and area of contact. Moreover, the fatal current differs widely from person to person. In fact there are people who can stand 250 V connected across their two thumbs while others would immediately die.

The electricity mains is a major source of danger, in many countries supplied at between 200 and 250 V. Generally one wire is connected to earth, called the *neutral* and contact with the neutral presents no danger, conversely the other *live* wire is at mains voltage relative to earth and will drive a current through the resistance of a human body in series with its connexion to earth as shown in Fig. 3.18 where R_b represents the resistance of the body between the point of contact with the live wire (e.g. hand) and the point or points of contact with the earth (e.g. through soles of feet or other hand). R_e represents the actual resistance of the connexion to earth which might be for example through plastic-soled shoes on a bedroom carpet (several megohms) or at the other extreme almost directly to earth through contact with a bathroom tap or from a power tool in the garden, both of the latter with a resistance of no more than a few ohms. In the first case there is no danger, probably not even a tingling sensation, in the second and third cases there is great risk. In between there are non-fatal shocks which not only live up to their name but may cause involuntary muscular movements which themselves give rise to injury or damage.

From this emerges one golden rule, treat all voltages over, say, 50 to 100 with great respect.

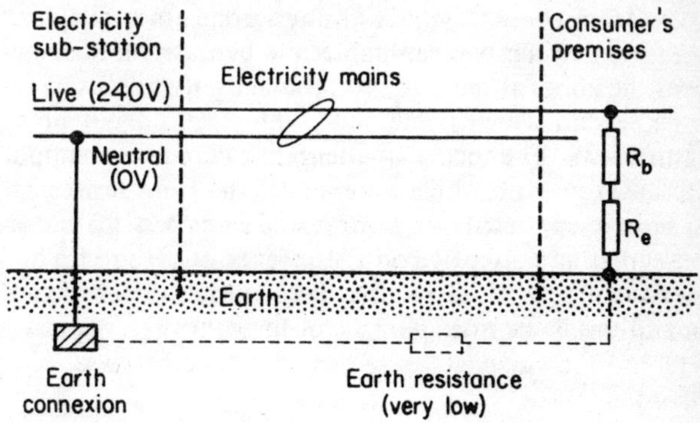

Fig. 3.18 Circuit for electric shock.

Fortunately modern technology has evolved the earth-leakage circuit-breaker to protect equipment and life from the electricity mains. This operates and disconnects the supply within about one-thirtieth of a second when a *leakage* current (not the main current) from the live wire reaches a value of only some 10 – 30 mA, thus providing maximum safety.

3.7 Direct and Alternating Currents

We need to pause at this point to look back on our progress and set the stage for the next move. The simple circuit so far considered was classed as such because it looked upon a circuit more or less as a loop of series resistances round which electrons were forced to flow by virtue of a difference of charge or e.m.f. somewhere within the loop. Parallel networks of resistances do not alter this thesis because as shown in Section 3.4.6 each can be reduced to a single series element. As long as the e.m.f. exists, then current continues to flow from the negative pole of the e.m.f. to the positive. An e.m.f. of this nature with fixed polarities gives rise to a *direct* (i.e. unidirectional) current. In electronics we have also to consider *alternating* currents and voltages which do not have fixed polarity but "swing" from one to the other at rates varying from about 25 times in one second for a telephone bell current which one can appreciate because the hammer strikes the gongs at that rate, to thousands of millions of times in one second which is beyond comprehension but not measurement. The theory of alternating current is so important that it needs a whole book and apart from resistors, two more components, *capacitors* and *inductors* feature in the alternating current circuit. However, these should be understood theoretically from a direct current point of view first and this is the main purpose of the next two chapters.

4. ELECTROSTATICS

This is the theory of static electricity. Although current flow is always electricity on the move, it is also possible to find electric charges which do not move (static). Typical examples of the generation of static electricity have already been given in Section 2.3, this chapter continues with the theoretical and practical aspects of electric charges. How a static charge can assist in a dynamic (active) process (electrons cannot do much if they stand still) will become evident later.

4.1 Capacitance

Just as a resistor is the physical form of resistance, so a capacitor is that of *capacitance* which is the capability of storing electric charge. A battery has one main function which is to push electrons out from its negative terminal while accepting a similar number in at its positive terminal. If it is therefore connected to a pair of wires which are unconnected at the remote ends as in Fig. 4.1(i), electrons will be forced along the upper wire and withdrawn from the lower one, leaving the top wire negatively charged and the bottom one positive. So a minute current flows but only for a very small fraction of a second until the potential on each wire is equal to that at the battery terminal, leaving no difference in potential to cause further electron flow.

Thus arises the condition where the wires are storing a quantity of electricity and if the battery were to be suddenly disconnected, the charge would remain.

4.1.1 The unit of capacitance

From the above it is evident that the quantity of electricity stored is proportional to the voltage applied for increasing the latter will cause a greater shift of electrons. There will be a quantity of electrons stored along the top wire and the same quantity removed from the bottom wire, we represent this total by Q coulombs and therefore $Q \propto V$[(A4.2.2)] where V is the applied voltage.

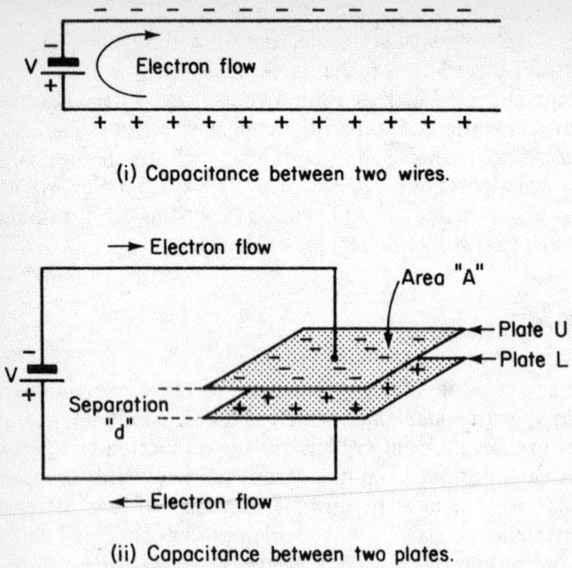

Fig. 4·1 Capacitance and opposing charges.

From this arises the definition of capacitance which is measured in *farads* (after Michael Faraday, the English chemist and physicist), one farad being that capacitance which when charged by one coulomb of electricity, produces a potential difference of one volt, thus

$$Q \text{ (coulombs)} = C \text{ (farads)} \times V \text{ (volts)}$$

or

$$C = \frac{Q}{V}, \qquad V = \frac{Q}{C}$$

In practice the farad which is large, is seldom used, more conveniently in electronics the microfarad ($\mu F = F \times 10^{-6}$) and picofarad ($pF = F \times 10^{-12}$) are used.

EXAMPLE:
A capacitor of 25 μF is given a charge of 0.001 coulombs. What will be the p.d. across the plates?

$$V = ? \qquad Q = 0.001\ C \qquad C = 25 \times 10^{-6} F$$

$$V = \frac{Q}{C} = \frac{0.001}{25 \times 10^{-6}} = \frac{0.001 \times 10^{6}}{25}$$

$$= \frac{0.001 \times \overset{4}{\cancel{100}} \times 10^{4}}{\cancel{25}} = \underline{40 \text{ volts}}$$

(split 10^6 into 100×10^4 so that the 25 will cancel).

4.2 The Simple Capacitor

If the wire size is increased, so also will be the capacitance just as a large jug has a greater capacity for holding liquid than a small one. The practical capacitor, which is a component deliberately designed to have a large capacity in a small space, exploits this by using instead flat metal plates as shown diagrammatically in Fig. 4.1(ii). Obviously the greater the area of the plates, the greater will be the capacitance and if the latter is denoted by C and the area of one plate by A, then $C \propto A$.

4.2.1 Electric fields

We must first understand the effect of the electric field which exists between any two bodies carrying different types of charge, as for example, the plates in Fig. 4.1(ii). In this case the upper plate U carries a negative charge, and the lower plate L a positive charge.

The accumulation of charge gives rise to an electric field. The proximity of the positive field of L to the negative charge on U and vice versa, would have the effect of reducing the potential difference between the plates were it not for the extra electrons which flow in order to maintain zero difference between each plate and the battery terminal to which it is connected. Thus the quantity Q of electricity stored has increased and since $C = Q/V$, if Q has increased while V remains constant, the capacitance C has increased.

The effect of the field of one plate on the charge held by the other increases as the plates get closer, that is, as the separation distance, d, decreases, hence

$$C \propto \frac{1}{d}$$

Combining this with the relationship of capacity to plate area:

$$C \propto \frac{A}{d}$$

4.2.2 *Adding a dielectric*

So far the plates have been considered as being separated in air. If some other insulating material is used between them it is found that the capacitance increases, which is fortunate because the material also prevents metallic contacts arising between them. When used in this way the material is called a *dielectric*. Again we must consider the electric field to find a satisfactory explanation. Instead of examining the effect of the field of one plate on the other, consider first the effect of both fields together on a single electron of the dielectric within the fields as in Fig. 4.2(i). The two fields obviously act together in this case, both exerting a force on the electron in the same direction, for example, downwards in the figure. The electron will not be released from its atom to move freely because this is just what insulating materials resist, actually

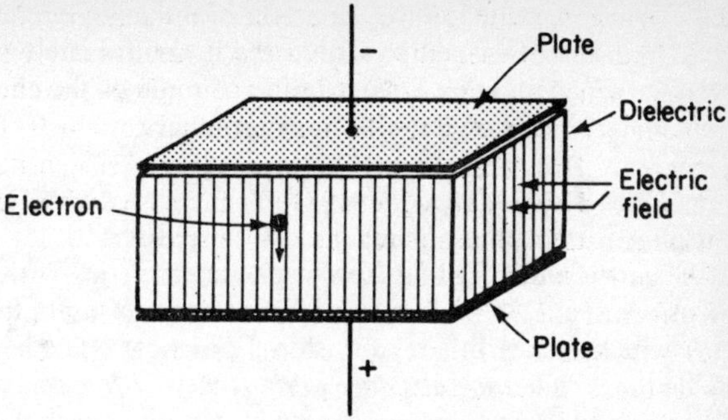

(i) Direction of force on electron in dielectric.

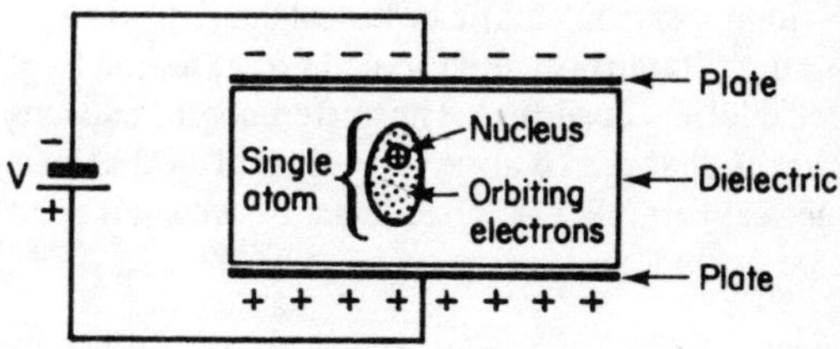

(ii) Atom distorted in electric field.

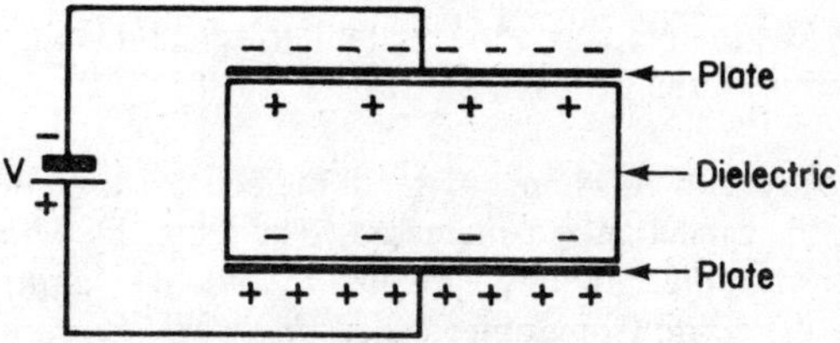

(iii) Shift of charge within the dielectric.

Fig. 4.2 Distortion of atoms by an electric field.

the atoms become distorted by the field as shown diagrammatically for one atom only in Fig. 4.2(ii). Here the electron orbits have been pulled in the direction of the positive plate, resulting in a shift of charge within the dielectric itself as shown in (iii). This causes cancellation of some of the charge on the plates, hence an increase in capacitance.

4.2.3 Calculation of capacitance

Returning to the simple parallel-plate capacitor with air as the dielectric and recalling how capacitance varies with dimensions, i.e. $C \propto A/d$, it is now necessary to insert the factor which makes this relationship a practical one[(A4.2.2)]. This factor is called the ***absolute permittivity of free space***, a fearsome sounding expression but what it really means to us is that this is a constant factor which may be inserted in the above relationship so that the capacity of a simple parallel-plate capacitor held in free space (i.e. a large vacuum containing no other near objects) can be calculated. The factor is 8.85×10^{-12} farads per metre and is represented by ϵ_0 (ϵ is the Greek letter, epsilon). The difference in capacity when free space is changed to air is very small (1.0006 times) and the practical formula for capacitance between air-spaced plates therefore becomes

$$C = \frac{A\,\epsilon_0}{d}\ \text{farads} = \frac{8.85\ A}{d} \times 10^{-12}\ \text{farads}$$

where A is the plate area in square metres and d is the separation between the plates in metres.

As already shown, when a material dielectric is used instead of air, the capacitance is increased. The factor by which this happens is represented by ϵ_r, called the *relative permittivity* of the dielectric (sometimes called *dielectric constant*) and

$$\epsilon_r = \frac{\text{capacitance with material as dielectric}}{\text{capacitance with air dielectric}}$$

making the formula finally

$$C = \frac{8.85\ A\ \epsilon_r}{d} \times 10^{-12} \text{ farads.}$$

Some typical values of ϵ_r for modern dielectrics are

Glass	Relative Permittivity 5 – 8
Mica	7
Paper	2.5 – 3.0
Paraffin Wax	2.2
Vulcanized Rubber	3.2
Ceramics	up to 1200
Polycarbonate	3.0
Polyester	3.1
Polypropylene	2.25
Polystyrene	2.6
Polythene	2.3

Working directly in pF the formula becomes

$$C = \frac{8.85\ A\ \epsilon_r}{d} \text{ pF.}$$

EXAMPLE:
A capacitor comprises two square plates of side 5 cms separated by glass 1 mm thick having a relative permittivity of 7.5. What is the capacitance?

$$A = (5 \times 10^{-2})^2 = 25 \times 10^{-4} \text{sq.m.}$$

$$d = 10^{-3}\text{m} \qquad \epsilon_r = 7.5$$

$$\text{Capacitance, } C = \frac{8.85 \times 25 \times 10^{-4} \times 7.5}{10^{-3}}$$

$$= 8.85 \times 25 \times 10^{-4} \times 7.5 \times 10^{3}$$

$$= 8.85 \times 25 \times 7.5 \times 10^{-1} = \underline{165.9 \text{ pF}}$$

EXAMPLE:
It is required to construct a capacitor having a value of 680 pF using two circular plates and a mica dielectric 0.1 mm thick of relative permittivity 7.0. What size plates are required?

Rearranging the formula:

$$C = \frac{8.85\,A\,\epsilon_r}{d} \text{ pF}$$

$$\therefore Cd = 8.85\,A\,\epsilon_r$$

$$\therefore A = \frac{Cd}{8.85\,\epsilon_r} \quad \text{where C is in pF.}$$

$C = 680$ pF $\quad d = 0.1 \times 10^{-3} (= 10^{-4})$m $\quad \epsilon_r = 7.0$

Then

$$A = \frac{680 \times 10^{-4}}{8.85 \times 7} \text{ sq.m} = 0.0010977 \text{ sq.m.}$$

$$= 0.0010977 \times 10^{6} \text{ sq.mm}^{(A5.3)} = 1097.7 \text{ sq.mm}$$

As the plates are to be circular, $A = \pi r^2$ where r is the radius of the plate.

$$\therefore r^2 = \frac{A}{\pi} \quad \text{and} \quad r = \sqrt{\frac{A}{\pi}}$$

$$\therefore r = \sqrt{\frac{1097.7}{\pi}} = \sqrt{349.41} = 18.69 \text{ mm}$$

∴ Circular plates should have a radius of 18.69 mm (or diameter 37.38 mm).

Capacitors need not be constructed from two plates only, the effective area is increased by interleaving plates as shown in Fig. 4.3 which shows the elements of a 5-plate capacitor, with its 4 dielectrics. Plates 1 and 2 with dielectric (i) form a simple capacitor. Equally plates 2 and 3 with dielectric (ii) form a second one of the same capacitance. The two sections together have twice the capacitance of one section and continuing this reasoning, the total capacitance of an interleaved system is equal to the capacitance of one section multiplied by the number of dielectrics or equally, (number of plates −1).

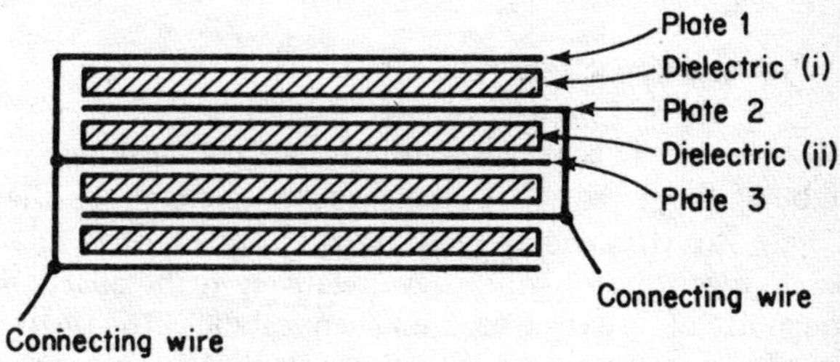

Fig. 4.3 Interleaved-plate capacitor.

Put in another way, Fig. 4.3 shows a capacitor equivalent to a simple parallel-plate capacitor of effective area A x n where n is the number of dielectric spaces, so giving:

$$C = \frac{8.85\,A\,\epsilon_r n}{d} \times 10^{-12}\ F$$

EXAMPLE:
A capacitor of 11 rectangular plates of dimensions 5 x 4 cms has a polycarbonate dielectric 0.2 mm thick. If the relative permittivity of the polycarbonate is 3.0, what is the capacitance?

$$A = 5 \times 4 = 20\text{ cm}^2 = 2 \times 10^{-3}\text{ m}^2$$

$$d = 0.2 \times 10^{-3}\text{ m} \qquad \epsilon_r = 3.0 \qquad n = 10$$

$$\text{Capacitance} = \frac{8.85 \times 2 \times 10^{-3} \times 3 \times 10}{0.2 \times 10^{-3}} \times 10^{-6}\ \mu\text{F}$$

$$= 8.85 \times 3 \times 10^{-4} = \underline{0.002655\ \mu\text{F}}$$

4.3 Practical Capacitors

Capacitors exist in several different forms, the most commonly used being the *fixed*, that is, the capacitance cannot be changed; these vary from small plastic-coated button or tubular types to *electrolytics* which have relatively high capacitance through use of an extremely thin chemical dielectric. Alternatively a capacitor may be *variable*, meaning that any value within a range can be chosen by rotation of a knob or screw.

4.3.1 Voltage rating

Fig. 4.2 shows at (i) that an electron within an electric field has a force acting on it, trying to move it towards the positive plate. At (ii) it is shown how an atom is distorted yet remains complete provided that the force on the electrons is insufficient to cause them to break away. As the voltage across the plates is increased however, the force on the electrons can ultimately dislodge them, thus constituting a normal electronic current. Because this flows through the high resistance of the insulating material, there may be a large heating effect under which the dielectric breaks down, that is the *breakdown voltage* is reached. Since the force exerted on an electron by a charge is greater as the distance between them decreases, it is evident that a thin dielectric will have a lower *voltage rating* (that is, a voltage sufficiently below the breakdown value for high reliability) than a thicker one, thus generally capacitors which have high capacitance through use of a thin dielectric, have low working voltages and vice versa.

4.3.2 Tolerance

As with resistors (Section 3.4.4), capacitors are quoted with a nominal value plus the tolerance. Standard tolerances are frequently ± 20% with ± 10%, ± 5% and even less for special purposes. For much design work where a high capacitance is required and it is not important if the value is exceeded, capacitors are often used with the tolerance very large on the positive side, e.g. –20%, +75% or even –10%, +100%. As with resistors and in fact all components, the greater the tolerance, the cheaper the item.

4.3.3 Construction

Because any two pieces of metal lying adjacent exhibit capacitance, practical capacitors may take on many forms, much depending on the value of capacitance required and the allowable size, e.g. from a few centimetres of insulated wire twisted together with a capacitance of a picofarad or so and a high working voltage to the aluminium electrolytic with a capacity of up to 0.1 F (10^{11} picofarads) in a can some 11 cms long but with a low working voltage of about 6.

Table 4.1 outlines therefore only a few of the many methods of construction. It is also clear from the list of values of ϵ_r given in Section 4.2.3 that several different types of plastic compound are available for dielectrics, some rejoicing in impressive names such as polyethylene terephthalate, polycarbonate, etc. To these must be added the "epoxy resins" and others. However, we will avoid where possible the use of the individual names as these are really the province of the chemical engineer, electronically our concern is with the characteristics of the material rather than its chemical composition, hence we continue with the general term "plastic".

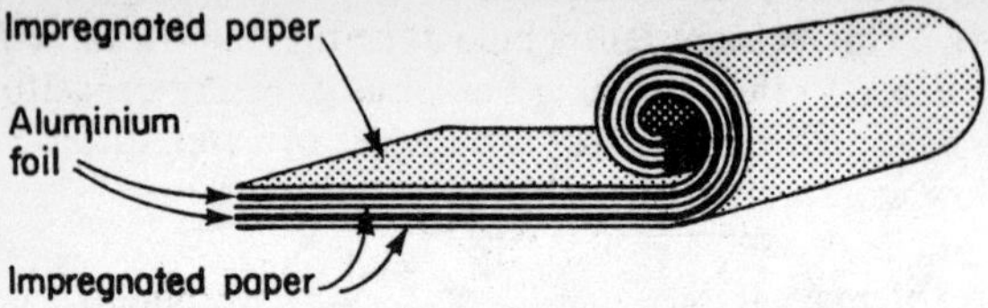

(a) Impregnated paper dielectric capacitor.

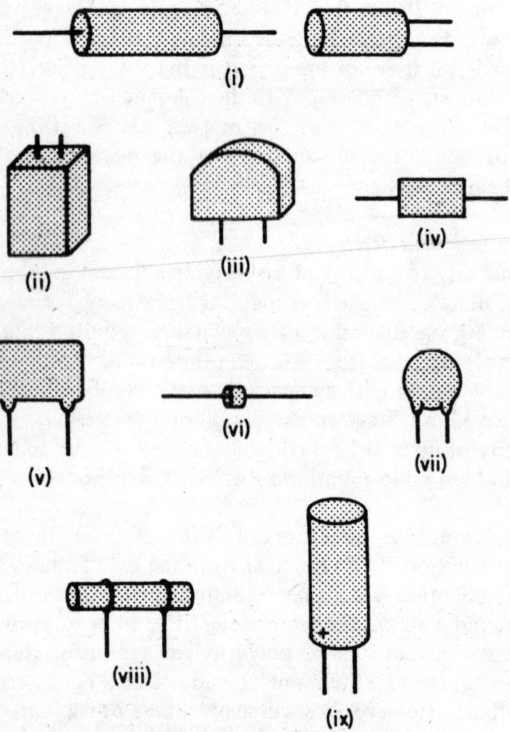

(b) Fixed capacitor shapes.

Fig. 4.4 Typical capacitor constructions.

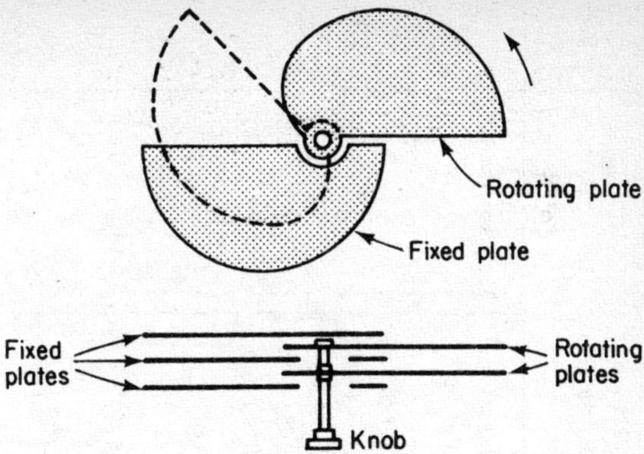

(c) Variable air dielectric capacitor.

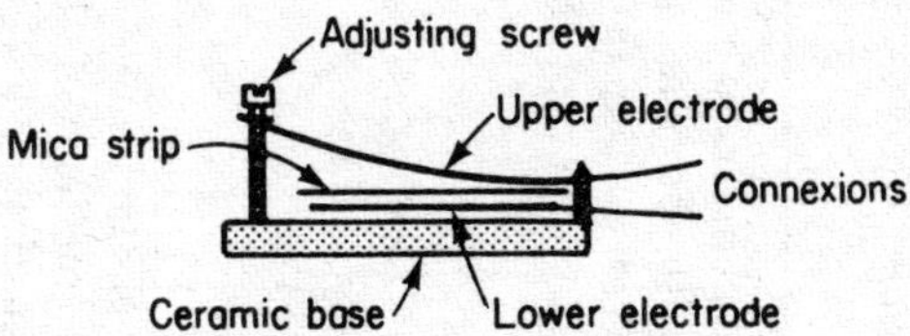

(d) Variable mica dielectric capacitor.

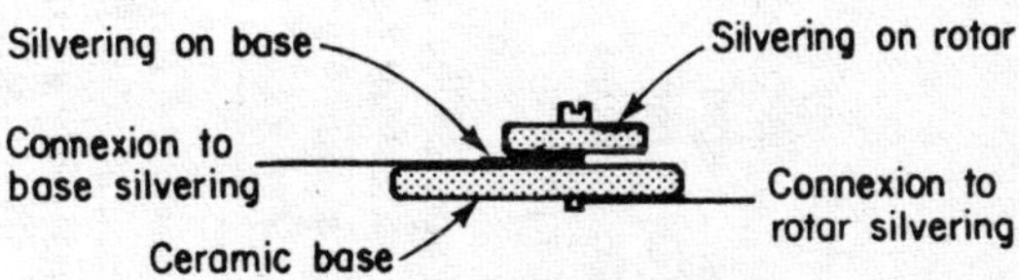

(Shown at minimum capacitance,
– maximum when screw rotated half a turn.)

(e) Rotary ceramic variable capacitor.

TABLE 4.1 **Typical Capacitors (Fig. 4.4 refers)**

Type (Dielectric)	*Construction*	*Appearance*	*Approx. Capacitance Range*	*Notes*
Impregnated Paper	Alternate layers of aluminium foil (plates) and paper impregnated with epoxy resin (dielectric). "Sandwich" rolled as in (a). Connexions made to foil by thin copper strips.	Tubular or Rectangular Can (b) – (i) or (ii)	up to a few μF	
Metallized Paper	Paper dielectric as above. Electrodes (plates) formed by thin layer of metal (e.g. aluminium) deposited on surface of paper	Flat pack (b) – (iii)	0.01 – 1.0 μF	
Metallized Plastic Film	As above, but using thin plastic film instead of paper. Various methods of assembly according to capacitance and type of plastic.	Can or flat pack (b) – (i), (ii), (iii), (v)	47 pF – 10 μF	

Mica	Very thin sheets of mica silvered on both sides, stacked and bolted together. The two silvered surfaces of adjacent dielectric sheets are in contact to form one electrode	Small flat pack (b) – (iv), (v)	Few pF to about 0.25 μF	
Ceramic	Silver electrodes formed onto ceramic disc, tube or plate. Multilayer ceramics (thin ceramic dielectrics with metal electrodes are stacked and then fired).	Bead, flat or tubular (b) – (vi), (vii), (viii)	0.5 pF – 0.05 μF 10 pF – 1.0 μF	Ceramic – from Greek keramos = pottery
Aluminium Electrolytic	Aluminium foils separated by gauze impregnated with an electrolyte. Very thin dielectric film of aluminium oxide formed on one foil.	Can (b) – (i), (ix)	1 – 100,000 μF	Correct polarity must be maintained across electrolytic capacitors when in use
Tantalum Electrolytic	As above but using one of the rarer metals, tantalum for higher values of capacitance in reduced size	From small plastic bead to tubular can e.g. (b) – (i) or (vi)	1 – 470 μF	

(table continued)

Type (Dielectric)	*Construction*	*Appearance*	*Approx. Capacitance Range*	*Notes*
Variable Air	Two sets of interleaved aluminium plates with variable overlap. Adjustment by knob at end of spindle fixed to rotating plates	See (c)	Various, typically over range 10 – 400 pF	Used for tuning radio receivers
Variable Mica	Metal strips separated by mica. Contact between upper strip and mica controlled by adjustable screw	See (d)	Single pair of plates, 2 – 40 pF. Multiplate, 150 – 2000 pF	
Rotary Ceramic	Partly silvered ceramic disc is rotated over silvering on a fixed base	See (e)	2 – 7 pF to 20 – 120 pF	

Again, in common with resistors, very small value capacitors may also be formed within "integrated circuits". The value is chosen at the circuit design stage and thereafter cannot be altered (or even seen!).

Apart from printing the actual value on a capacitor, a colour-dot system may be used, nearly all such systems are constructed around the 3-dot colour code which is given in Appendix 10.

4.4 Capacitors in Series and Parallel

We now move from construction sketches as in Fig. 4.4 to consideration of capacitance values, so sketches give way to the circuit symbol[(A7)] for a fixed capacitor as shown in Fig. 4.5. Here, let the two capacitors be of any type but having capacitances C_1 and C_2 (F, μF, pF, it does not matter at this stage). They can be connected together as with resistors, in either of two ways, series or parallel. We examine parallel first because it is a little more straightforward.

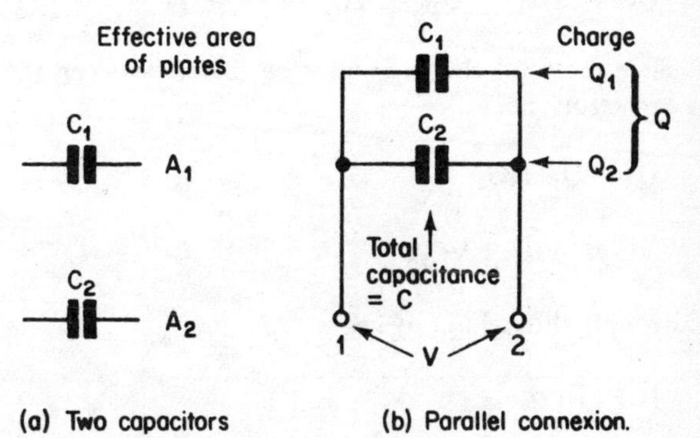

Fig. 4.5 Capacitors connected in parallel.

Parallel:
Connecting both plates (or sets of plates) of C_1 to the plates of C_2 as shown diagrammatically in (b) is simply equivalent to increasing the total plate area to $A_1 + A_2$, and therefore because capacitance varies directly as the effective area of plates, the total capacitance becomes $C_1 + C_2$ so if C is used to denote this total capacitance,

$$C = C_1 + C_2$$

and similarly it can be shown that

$$\underline{C = C_1 + C_2 + C_3 + \ldots}$$

where any number of capacitors are connected in parallel.

This can be considered a little more scientifically as follows:

Let a voltage, V, be applied across a parallel combination as shown in Fig. 4.5(b). This voltage is applied equally to both capacitors. Let the charge on C_1 become Q_1, on C_2, Q_2 and the charge on the combination, Q.

Now $Q_1 = VC_1 \quad Q_2 = VC_2 \quad Q = VC$

and since the total charge Q must be shared between the two capacitors, i.e.

$$Q = Q_1 + Q_2$$

$$\therefore \quad VC = VC_1 + VC_2$$

and dividing throughout by V,

$$C = C_1 + C_2$$

and similarly

$$\underline{C = C_1 + C_2 + C_3 + \ldots}$$

Series:
Consider next two capacitors C_1 and C_2 connected in series as in Fig. 4.6 and again let a voltage, V be applied to the combination. As above, it is necessary to consider the charges set up by the movement of electrons when V is applied. If, for example, V has polarity as shown, electrons will flow from the right-hand (r.h.) plate of C_2 to terminal 2 which V maintains positive and equally the excess of electrons caused by V at terminal 1 creates a flow of current into the l.h. plate of C_1 and these two movements of charge are equal. Considering C_1 only, the r.h. plate will be influenced by the electrostatic field through the dielectric so driving electrons from this plate, whereas the opposite occurs at the l.h. plate of C_2, the net effect being a movement of electrons from C_1 to C_2, that is, a current flows even though this path is actually insulated from the voltage V, by the dielectrics. The overall effect is that a momentary flow of electrons occurs from terminal 1 to 2 due to the application of V. Both capacitors receive the same level of charge, let this be Q.

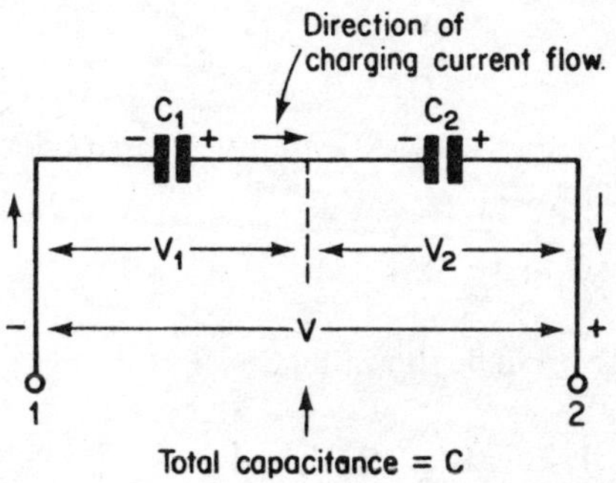

Fig. 4.6 Capacitors connected in series.

Now, if at any instant a voltage V_1 exists across C_1 and V_2 across C_2, then

$$V = V_1 + V_2$$

and since the charge on both C_1 and C_2 is Q

$$V_1 = \frac{Q}{C_1} \qquad V_2 = \frac{Q}{C_2}$$

$$\therefore \quad V = \frac{Q}{C_1} + \frac{Q}{C_2}$$

and dividing both sides by Q,

$$\frac{V}{Q} = \frac{1}{C_1} + \frac{1}{C_2}$$

and since

$$\frac{Q}{V} = C, \qquad \frac{V}{Q} = \frac{1}{C}$$

$$\therefore \quad \frac{1}{C} = \frac{1}{C_1} + \frac{1}{C_2}$$

and similarly it can be shown that

$$\underline{\frac{1}{C} = \frac{1}{C_1} + \frac{1}{C_2} + \frac{1}{C_3} + \ldots}$$

a formula of similar form to that for resistances in *parallel.*

EXAMPLE:
Three capacitors of 1 μF, 0.47 μF and 33,000 pF are connected in parallel. What is the combined capacitance?

$$C = C_1 + C_2 + C_3$$

$$= 1.0 + 0.47 + (33{,}000 \times 10^{-6})\ \mu F$$

$$= 1.47 + 0.033\ \mu F = \underline{1.503\ \mu F}$$

EXAMPLE:
Two capacitors of 1 and 3 μF are connected in series. What is the combined capacitance and what voltage will exist across each if 300 V is applied to the combination?

$$\frac{1}{C} = \frac{1}{1} + \frac{1}{3}\ \mu F = \frac{4}{3}$$

where $C_1 = 1\ \mu F$, $C_2 = 3\ \mu F$

$$\therefore C = \frac{3}{4} = 0.75\ \mu F$$

[or slightly more quickly as we found with resistors in parallel (Section 3.4.5)

$$C = \frac{C_1 C_2}{C_1 + C_2} = \frac{1 \times 3}{1 + 3} = \frac{3}{4}]$$

The charge Q is the same on both capacitors and is equal to that for a capacitance of 0.75 μF with 300 V applied, i.e.

$$Q = 0.75 \times 10^{-6} \times 300 \text{ coulombs} = 225 \times 10^{-6}\ C$$

∴ Voltage across 1 μF capacitor

$$= \frac{Q}{C_1} = \frac{225 \times 10^{-6}}{10^{-6}} = \underline{225\ \text{V}}$$

∴ Voltage across 3 μF capacitor

$$= \frac{Q}{C_2} = \frac{225 \times 10^{-6}}{3 \times 10^{-6}} = \underline{75\ \text{V}}$$

thus demonstrating that the higher voltage is developed across the lower capacitance.

5. ELECTROMAGNETISM

Nature gave us yet another powerful but unseen force, that of *magnetism*. An oxide of iron, known as magnetite was first discovered in the Middle East as having the property of attracting pieces of iron and of swinging round to point North and South when freely suspended. The use of the suspended magnet as a compass by early mariners led to the name of "lodestone" (lode + stone, literally "way-stone" – by which mariners found their way). It was subsequently discovered that a small, elongated piece of iron could be magnetized by stroking it with lodestone, it too would then exhibit magnetic properties concentrated mainly near the two ends. Suspending this piece of iron again resulted in one of the ends always pointing to one particular pole of the Earth from which it was clear that the magnetism at each end of the magnet was different. The *poles* of the magnet as they are called, have been labelled *North* and *South* as is that huge magnet, the Earth. Both poles of a magnet will attract iron equally well but if two such magnets are brought together "like poles repel, unlike poles attract", a rule very similar to that for electric charges, applies. A compass needle is simply a very light bar magnet on a centre pivot as friction-free as possible, one of its poles will swing round until it points to the North Magnetic Pole of the Earth, thus anywhere on Earth the direction of its poles can be determined. Note that we speak of the Earth's magnetic poles, these are in fact slightly displaced from the *true* North and South Poles.

5.1 Magnetization

We found earlier that both electrical charge and magnetism are associated with atoms and now it seems that they both have polarities which obey the same rule "like repel, unlike attract". Further progress will show that electric fields and magnetic fields are intimately associated and to understand this relationship a little more clearly let us return to the atom once more.

The electron spins on its own axis and orbits the nucleus just as the Earth does. It carries an electric charge and moving charges have associated with them magnetic fields (Section 2.2). Thus each electron, because it is continually spinning is equivalent to a tiny magnet and Nature arranges things so that the total magnetic effect of all the electrons of an atom is usually negligible through mutual cancellations. However, in some materials this is not so and the whole atom then acts as a small magnet itself, having its own North and South poles, such materials are for instance, iron, steel, nickel and cobalt.

Just as positive and negative charges on the plates of a capacitor have an electric field between them which exerts a force on any charge within it, so two magnetic poles have a magnetic field between them which is capable of exerting a force on another magnetic pole. There is one main difference, electric charges exist on their own but magnetic poles do not, we cannot have a single N or S pole, only the pair as part of a complete magnet.

Returning to the atom of, for example, iron, which by electron theory we can represent as a miniature magnet, if the atom is within a magnetic field then there will be a force pulling the N pole of the atom towards the S pole of the external field and similarly with the opposite pole. If the field is of sufficient strength the atom will align itself with it. In a "hard" metal such as steel the atoms remain aligned even when the external field is removed, with all atom N poles pointing in one direction and S poles in the opposite, the steel exhibits *residual* magnetism, that is, it has itself become a *permanent* magnet. In "soft" iron, however, the atoms revert to their normal random positions when the field is removed and very little or no magnetism remains.

5.1.1 Magnetic fields

Confusion easily arises over magnet polarities when the Earth's magnetism is considered. The end of a compass-needle or suspended magnet which points north must be of opposite magnetic polarity to that at the North Pole of the Earth so we call the end "North-seeking" to be technically correct, but frequently label it "North" or just plain "N". The opposite pole is "South-seeking" or "S", which in effect means that we label all our magnets oppositely to the way we label the Earth. In magnetic field diagrams on which a magnetic field is shown graphically by lines drawn to represent *magnetic flux*, the direction of the flux has been chosen as running from North-seeking to South-seeking, on such a diagram the Earth's field therefore runs South pole to North.

The following section includes a magnetic field diagram which will bring the points made in this and the preceding sections into focus.

5.1.2 Magnetic induction

Induction of a magnetic flux into a normally unmagnetized piece of metal such as soft iron is most easily explained by means of a magnetic field diagram as in Fig. 5.1. N and S are opposite poles of two bar magnets (or a single "horseshoe" magnet) as shown in (i). When only air separates the two poles the magnetic field is represented by the dotted lines running from N to S as shown. Insertion of the small piece of metal between the poles as in (ii) shows a crowding of the flux lines through the metal, inducing poles in it by alignment of the atoms and the direction of the lines from N to S is unchanged. The greater number of lines in (ii) shows that the flux has increased. The magnetism now exhibited by the soft iron is said to be the result of *magnetic induction*.

Magnetic induction is a feature of transformers, electric motors, generators and electromagnetic devices generally.

5.2 The Magnetic Circuit

There are certain features of the magnetic circuit which are not unlike those of the electric circuit in which a driving force (e.m.f.) produces some sort of output (current) against resistance, linked by the relationship that the output is inversely proportional to the resistance but proportional to

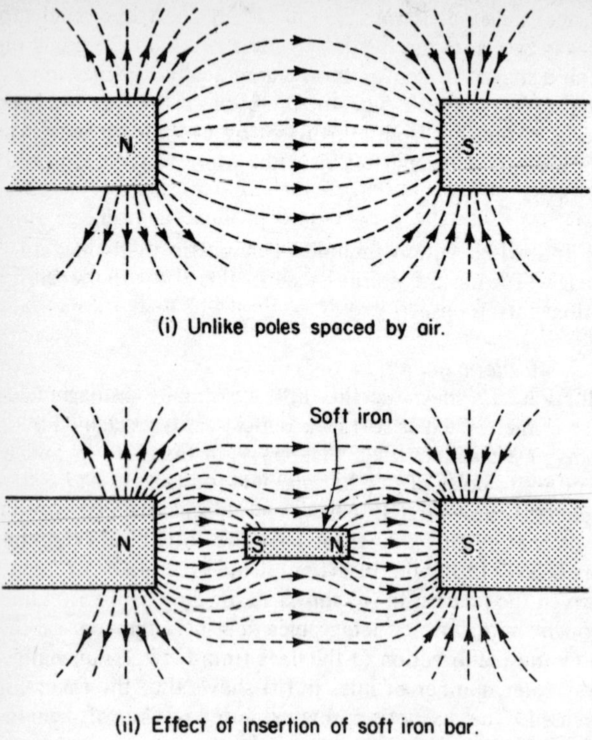

Fig. 5.1 Magnetic induction.

the driving force. The same holds good in the magnetic circuit where the driving force is called the *magnetomotive force* (m.m.f.), the output it achieves is the *magnetic flux*, operating against the *reluctance* of the magnetic circuit, giving a general expression:

$$\text{Total flux} = \frac{\text{magnetomotive force}}{\text{total reluctance}}$$

of similar form to $I = \frac{E}{R}$ for the electric circuit.

5.2.1 Magnetomotive force

It has been stated that moving charges carry with them magnetic fields, hence current flow in a wire, which is the result of an enormous number of small charges in motion in one direction might be expected to produce a magnetic flux provided that the individual fields add up in some way. This can be shown to be so and the direction of the magnetic flux bears an easily determinable relationship to the direction of current.

We have to be very careful again about electron flow and conventional current flow (Section 1.2.4) for the rules which were derived by early scientists were all based on the latter. To avoid confusion we will again ignore the outdated conventional current flow and the rules have been changed accordingly. The point is made in case the reader finds rules elsewhere which seem to tell the opposite.

If a wire carrying a current is held vertically through a piece of card as shown in Fig. 5.2(i), the magnetic flux would encircle the wire horizontally as shown and its direction could be determined by moving a small (pocket) compass on the card and noting at each move the direction in which the needle pointed (i.e. the N-seeking end, usually coloured – see Fig. 5.2(iv)). A current in the wire of several amperes would be required otherwise the Earth's magnetic field would tend to take over.

When electronic currents and their associated magnetic fields are shown diagrammatically we sometimes need to view the wire end-on, representing it by a circle. The current must then be flowing downwards into the paper or upwards out of it and this is unforgettably depicted by imagining an arrow to be within the wire and pointing in the direction of current flow as shown in Fig. 5.2(ii).

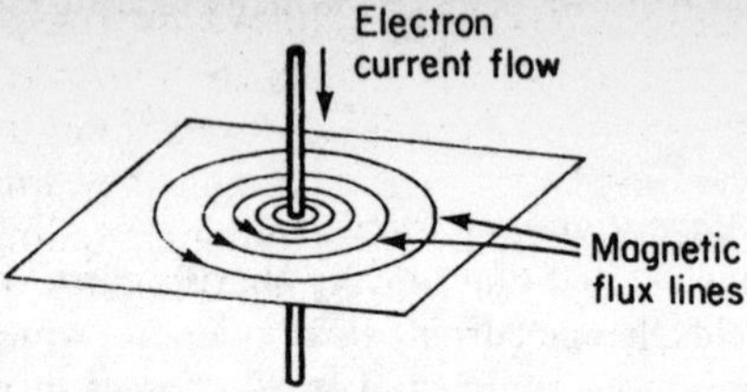

(i) Pictorial representation of magnetic field surrounding a wire.

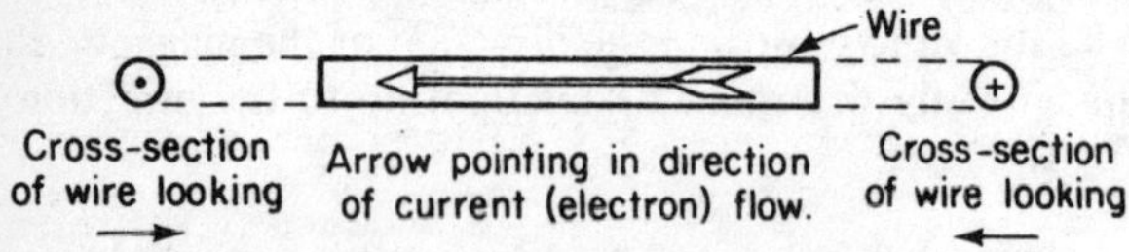

⊙ Depicts current flowing up out of paper.

⊕ Depicts current flowing down through paper.

(ii) Diagrammatic illustration of current direction.

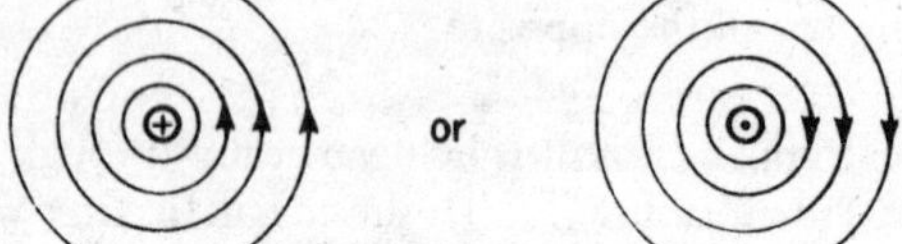

(iii) Diagrammatic form of (i).

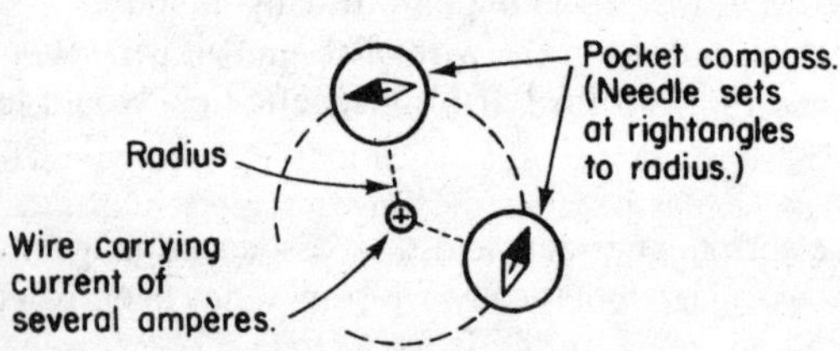

(iv) Method of plotting magnetic field using small compass.

Fig. 5.2 Magnetic flux diagrams.

Some help may be needed to remember the flux direction as shown in Fig. 5.2(i) and a useful way is to recall the similarity between current and water flow, from this we note that when water flows out of a tap, i.e. current flows downwards (or away from the observer), the tap must be turned anti-clockwise (direction of flux). The alternative follows automatically i.e. current upwards (towards the observer), flux clockwise.

An equally useful reminder is the "Left-Hand Rule". The wire is imagined to be grasped in the left hand (by wrapping the four fingers round it) with the thumb extended along the wire pointing in the direction of the current. The fingers are then pointing in the direction of the flux.

This leads directly to the very important concept that a coil of wire carrying a current acts as a magnet, and poles are formed, whereas with a straight wire they are not. This can be seen from Fig. 5.3 which shows that all the flux through the centre of the coil is in the same direction, the coil therefore acts as a magnet with N and S poles as shown for the particular current direction. The flux outside the magnet travels from N to S and within the magnet from S to N. The magnetic circuit is through the centre of the coil returning via the total path outside and around the coil.

Another Left-Hand Rule helps to determine directions. In this case the indications given by fingers and thumb are reversed compared with the use of the rule for a wire. Imagine the left hand to be grasping the coil with the fingers pointing in the direction of the current, the extended thumb then points in the direction of the flux, i.e. towards the N pole of the coil. There should be no confusion in having two different Left-Hand Rules because the fingers represent the circular component (flux in the case of the wire, current for the coil) in both cases.

The magnetic flux produced in this *air-cored* coil depends on two factors, the current and the number of turns of wire, from which

$$\text{magnetomotive force (m.m.f.)} = N \times I \text{ ampere-turns (At)}$$

where N = number of turns and I = current in amperes.

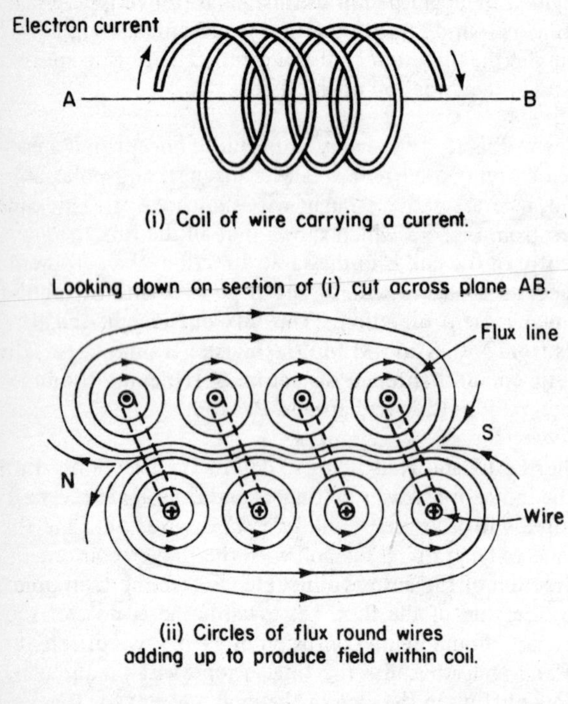

Fig. 5.3 Magnetic flux through coil carrying a current.

5.2.2 Reluctance

The parallel of electrical resistance is magnetic reluctance. Section 5.1.2 shows that the insertion of soft iron between two unlike poles increases the flux, it can equally be said that the soft iron has decreased the reluctance. Reluctance is denoted by the letter S and a mathematical proof of how S varies with the dimensions of the iron is unnecessary because it is fairly obvious that the longer the piece of iron, the greater will be its reluctance, also the greater the cross-sectional area, the lower it will be, exactly as in the case of resistance (remember $R = (\rho\ell/a)$)

$$\therefore S \propto \frac{\ell}{a}$$

where ℓ is the length of the magnetic path under consideration and a is its cross-sectional area.

The third variable is in the material itself. As with capacitors where we find the absolute permittivity of free space to obtain a practical relationship and follow this with a value for relative permittivity for a particular dielectric other than air, so in the magnetic case we do the same except that the term is now *permeability* instead of permittivity.

The *absolute permeability of free space* is designated by μ_0 and has a value of $4\pi \times 10^{-7}$.

Again, when air is substituted for free space, the change in μ is negligible so in practice

$$S = \frac{\ell}{\mu_0 a} = \frac{\ell}{4\pi \times 10^{-7} a} = \frac{10^7 \ell}{4\pi a}$$

for a flux path consisting solely of air.

The unit of magnetic flux is the Weber (symbol Wb, after Wilhelm Eduard Weber, a German physicist – correctly pronounced "vayber") and the standard definition of the Weber is written round the utilization of a magnetic field. At this point we have only considered the creation of one therefore it is sensible for the definition to be given later when the use of magnetic fields is examined.

Relative permeability, designated μ_r is the ratio of the density of the magnetic flux produced in a material compared with that produced in air by the same magnetizing force. Therefore, the reluctance of a path containing some material is

$$S = \frac{\ell}{\mu_0 \mu_r a} = \frac{10^7 \ell}{4\pi \mu_r a} \text{ ampere-turns/Weber}$$

The unit ampere-turns/Weber reminds us that reluctance is a measure of the amount of force required (ampere-turns) to create a certain value of flux (Wb).

[So many new terms introduced in only a few pages may seem confusing to the reader who meets them for the first time, but it is necessary to lay the foundations for theory which follows, much of this will be clearer as the practical outcome becomes evident.]

Reluctances in series are similar to resistances in series, and are therefore additive, hence for a magnetic path containing two or more reluctances in series:

$$S = \frac{10^7}{4\pi}\left(\frac{\ell_1}{\mu_{r_1} a_1} + \frac{\ell_2}{\mu_{r_2} a_2} + \ldots\right) \text{At/Wb}$$

where $\mu_{r_1}, \mu_{r_2} \ldots$ are the relative permeabilities of the several materials making up the complete magnetic path.

$\left.\begin{matrix} \ell_1, \ell_2 \ldots \\ a_1, a_2 \ldots \end{matrix}\right\}$ are the lengths and areas of cross-section of the materials.

5.2.3 Calculation of total flux

From the expressions for magnetomotive force and reluctance, the total flux in Webers can be calculated:

$$\text{Total flux} = \frac{\text{magnetomotive force}}{\text{total reluctance}}$$

In electronic shorthand, flux is denoted by Φ (the Greek capital letter, Phi), m.m.f. by F and as already shown, reluctance by S

$$\therefore \quad \Phi = \frac{F}{S}, \qquad F = \Phi.S \qquad S = \frac{F}{\Phi}$$

In practice the magnetic field calculations for an open coil as in Fig. 5.3 are complicated for several reasons, especially that the m.m.f. is not constant around the magnetic circuit and that the length and area of cross-section of the air-path outside the coil are not easily measured. Fortunately most applications employ a magnetic circuit which is almost completely through magnetic material, often with a small air-gap, for example, in transformers, generators, motors, loudspeakers, relays, magnetic recording heads. However, even in these cases the relative permeability is not constant, its value varies with the flux density which is often what we are trying to calculate, hence simple magnetic circuit calculations lead only to approximate answers. Nevertheless a few examples follow mainly to help in developing a feeling for the electromagnetic circuit of devices such as those mentioned rather than in expectation that the reader will be called upon or need to carry out such calculations, this being reserved for the few concerned with design.

The essential points are:

(i) it is electronic current flow from which the field arises;
(ii) in most applications the current flows through a coil of wire, each turn adding its quota to the magnetic flux;

(iii) the flux path follows a closed circuit which may consist of several sections having different reluctances;

(iv) lower reluctance allows greater flux, air and non-magnetic materials have highest reluctance.

EXAMPLE:

A coil of 1000 turns is wound on a steel ring (relative permeability 2000), the ring having a mean radius of 10 cm and cross-sectional area of 2.5 sq. cm. If a current of 5 amperes is passed through the coil, what is the total flux in the ring?

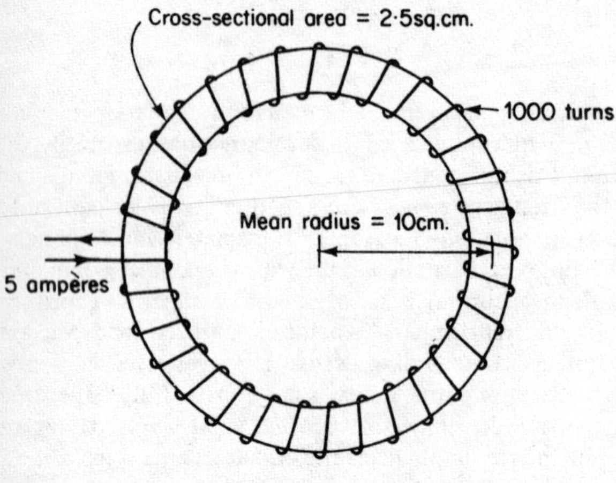

Fig. 5.4 Coil wound on steel ring.

The ring is shown diagrammatically in Fig. 5.4. Bring all dimensions to metres and square metres.

$N = 1000 \qquad I = 5A$

$\ell = 2\pi \times \text{radius} = 2\pi \times 10 \text{ cms} = 20\pi \times 10^{-2} \text{ m}$

$\mu_r = 2000 \qquad a = 2.5 \text{ sq. cm} = 2.5 \times 10^{-4} \text{ sq. m.}$

Then,

$$\text{m.m.f., } F = NI = 1000 \times 5$$

$$= 5000 \text{ ampere-turns (At)}$$

Reluctance,

$$S = \frac{10^7 \ell}{4\pi\mu_r a} = \frac{10^7 \times 20\pi \times 10^{-2}}{4\pi \times 2000 \times 2.5 \times 10^{-4}}$$

$$= 10^6 \text{ ampere-turns/Weber (At/Wb)}$$

Total flux,

$$\Phi = \frac{F}{S} = \frac{5000}{10^6} \text{ Wb} = \frac{5}{10^3} \text{ Wb} = \underline{5.0 \text{ mWb}}$$

EXAMPLE:
A saw cut 1 mm wide is made in the steel ring of the previous example. Again, calculate the total flux in the ring.

Let $\ell_1 a_1 \mu_{r_1}$ refer to the steel ring, $\ell_2 a_2 \mu_{r_2}$ refer to the air-gap.

$$F = 5000 \text{ At}$$

$\ell_1 = 20\pi \times 10^{-2}$ m (we can neglect the tiny reduction of 1 mm)

$a_1 = 2.5 \times 10^{-4}$ sq. m.

$\mu_{r_1} = 2000 \qquad \mu_{r_2} = 1$ (air or vacuum)

$\ell_2 = 1 \times 10^{-3}$ m

$a_2 = 2.5 \times 10^{-4}$ sq. m

Note that for a_2 we assume that all the flux crosses straight across the gap, that is, we ignore the small effect of *fringing* in which some of the flux bulges outwards from the gap as shown dotted in Fig. 5.5. With long air-gaps the effect of fringing may not be negligible.

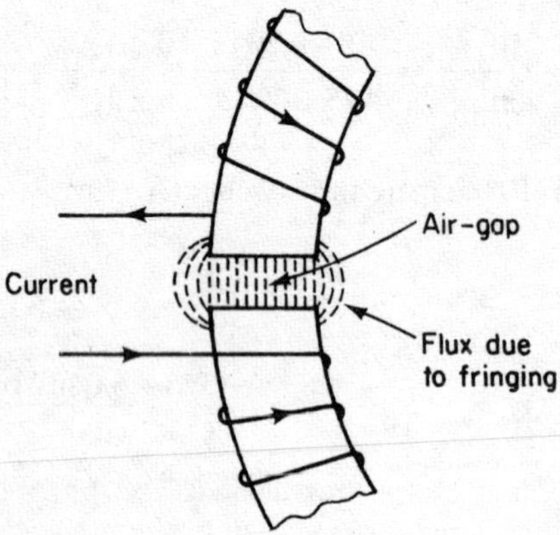

Fig. 5.5 Air-gap in steel ring.

The total reluctance (the sum of the reluctances of the steel and of the air-gap) will be increased because part of the magnetic circuit now has a lower permeability. For illustration, consider the two paths separately:

Reluctance of steel ring =

$$\frac{10^7 \ell_1}{4\pi\mu_{r_1} a_1} = 10^6 \text{ At/Wb (as before)}$$

$$\text{Reluctance of air-gap} = \frac{10^7 \ell_2}{4\pi\mu_{r_2} a_2} = \frac{10^7 \times 1 \times 10^{-3}}{4\pi \times 1 \times 2.5 \times 10^{-4}}$$

$$= \frac{10^4}{\pi \times 10^{-3}} = \frac{10^7}{\pi} = 3.183 \times 10^6 \text{ At/Wb}$$

(the 1 mm air-gap has over 3 times the reluctance of the whole steel ring).

∴ Total reluctance,

$$S = 10^6 + (3.183 \times 10^6) = 4.183 \times 10^6 \text{ At/Wb}$$

Then, total flux

$$\Phi = \frac{F}{S} = \frac{5000}{4.183 \times 10^6} \text{ Wb}$$

$$= \frac{5}{4.183} \text{ mWb} = \underline{1.195 \text{ mWb}}$$

These two examples together demonstrate the effect of small air-gaps in magnetic circuits. Fig. 5.6 shows some practical examples of magnetic circuits in which air-gaps either cannot be avoided and are therefore reduced to the minimum possible, (i) and (ii) or are used to deliberately create a high reluctance as in (iii).

A little revision may not come amiss on magnetic flux direction. Consider Fig. 5.5. With no air-gap the flux is continuous round the ring and poles are not formed but when an air-gap exists it may be considered that poles arise having a m.m.f. which drives the flux across the gap. If the current flows in the direction shown in Fig. 5.5, the flux lines will be set up inside the ring in an anti-clockwise direction so crossing the air-gap downwards in the figure, that is, forming a N-pole above the gap and S-pole below.

5.2.4 Flux density and magnetizing force

The total flux Φ is not truly a measure of the strength of a field because no account has been taken of the cross-

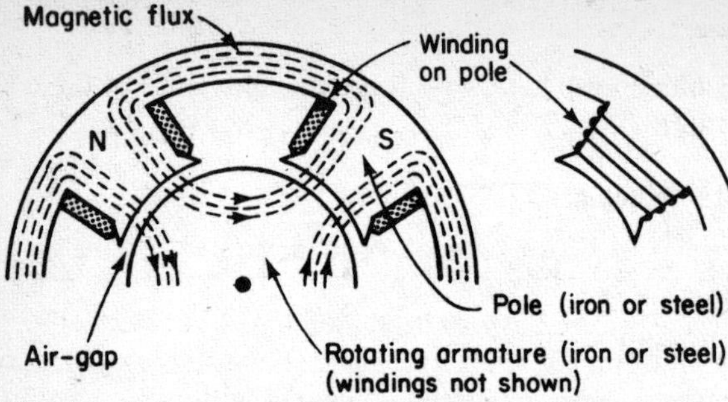

(i) Part of magnetic circuit of dynamo or motor.

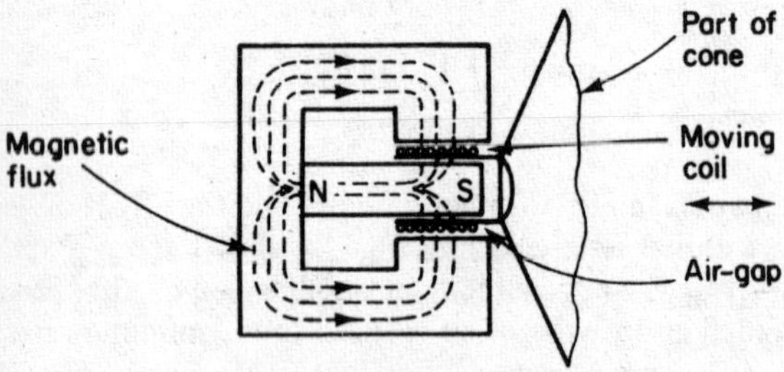

(ii) Cross-section of moving coil loudspeaker.

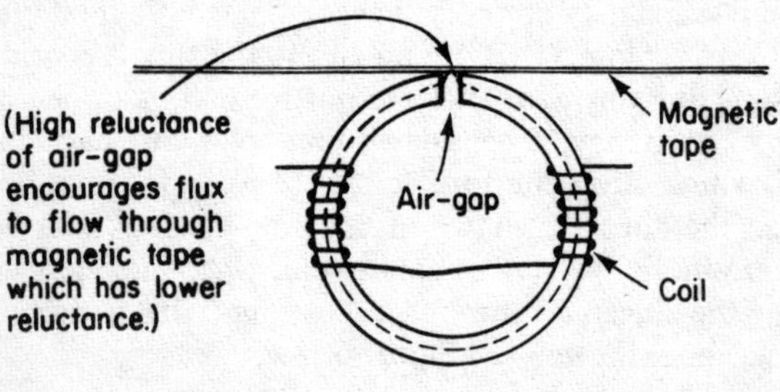

(iii) Magnetic record or playback head.

Fig. 5.6 Magnetic circuits with air-gaps.

sectional area through which it operates, a flux passing across a large area will be less dense than when it is confined within a smaller area. *Flux density* is given the symbol B and

$$B = \frac{\Phi}{a}\ \text{Wb/m}^2 = \frac{\Phi}{a}\ \text{tesla (T)} \quad (1\text{T} = 1\text{Wb/m}^2)$$

where Φ is in Webers, and a is in sq. metres.

Magnetizing force has also been mentioned but not defined. For a given m.m.f. the actual magnetizing force will be less when the flux lines are long compared with when they are shorter. A standard unit to express magnetizing force is therefore m.m.f. per unit length, it is given the symbol H and is measured in ampere-turns per metre, i.e.

$$H = \frac{F}{\ell}\ \text{ampere-turns/metre}$$

where F is in ampere-turns and ℓ is in metres.

Magnetizing force is also known as the *magnetic field strength* or *magnetic field intensity*.

Of interest, especially to those who may ultimately need to study characteristics of magnetic materials (for example for electromagnetic switching or information storage such as used in computers), is the relationship between B and H.

As shown previously

$$\Phi = \frac{F}{S} \quad \text{and} \quad S = \frac{\ell}{\mu_0 \mu_r a}$$

$$\therefore \quad \Phi = \frac{F}{\dfrac{\ell}{\mu_0 \mu_r a}} = \mu_0 \mu_r a \frac{F}{\ell}$$

$$\therefore \quad \frac{\Phi}{a} \text{ which equals } B = \mu_0\mu_r \frac{F}{\ell} = \mu_0\mu_r H$$

$$\therefore \quad \frac{B}{H} = \mu_0\mu_r \quad (\mu_0 \text{ is the constant } 4\pi \times 10^{-7})$$

therefore the ratio of the flux density to the magnetizing force is directly related to the permeability of a material. μ_0 is constant therefore $B \propto H.\mu_r$ indicating that as H increases, B also increases. However, in practice, at the higher flux densities (B), the relative permeability (μ_r) decreases, thus increase in B as H increases is offset by a fall in μ_r and eventually "magnetic saturation" is reached, that is, there is a limit to the flux density for any particular material.

5.3 Electromagnetic Induction

Although Michael Faraday's name is linked with the capacitor (Section 4.1.1), he is perhaps better known for his early work on electromagnetic induction for he discovered that an e.m.f. was produced in a wire when it was moved quickly across the space between a pair of magnet poles. Further work showed that a stationary wire with a magnetic flux moving across it also generated an e.m.f. In fact any variation of flux does so even when a flux is being set up or is collapsing as happens when a current is switched on or off in a coil of wire. When the motion is such that the wire travels along the lines of flux (that is, no flux is being "cut"), no e.m.f. is induced.

The magnitude of the e.m.f. depends on (i) the rate of cutting the flux, that is, in the case of the moving conductor, its speed, and (ii) on the strength of the flux itself.

That this does happen we may appreciate from our knowledge that the electron may be considered as a tiny magnet (Section 5.1). Consequently an external magnetic field sweeping across the conductor will accelerate or "kick" electrons into one direction or the other along the wire depending on the magnetic polarities of both the field and the electrons. The faster the external field changes, the greater the kick, in a way demonstrated by the distance a cricket ball will move according to the swing or rate of movement of the bat. A physicist would be unhappy with this explanation but it serves the purpose if it is not too far from the truth and helps us to remember the fundamental theory.

Pictorially the production of an induced e.m.f. is shown in Fig. 5.7, (i) shows a conductor moving across a magnetic field, (ii) shows how a magnetic flux associated with a bar-magnet produces a greater e.m.f. in a coil of wire, each turn of wire adding to the total effect, and (iii) shows the bar-magnet replaced by a coil of wire carrying a current. If this coil is moved in or out of the larger coil an e.m.f. is set up, also one will arise when the current is first switched on, and again, but in the opposite direction, when it is switched off – in each case the magnetic flux is changing. No e.m.f. is induced in (i), (ii) or (iii) when nothing changes. In (iii) we meet basic *transformer* action.

It is now convenient to consider the definition of the unit of magnetic flux, the Weber which had to be postponed in Section 5.2.2. The Weber is defined as the flux which, linking a circuit of one turn produces in it an e.m f. of one volt as it is reduced to zero at a uniform rate in one second, thus

$$\text{e.m.f. (volts)} = \frac{\text{total flux cut (Wb)}}{\text{time (secs)}}$$

If the length of the conductor within the field of Fig. 5.7(i) is ℓ metres and the flux density of the field is B Wb/m^2, then if the conductor moves downwards through the field a distance d metres in t seconds, the total area of flux cut is $\ell \times d$ sq. m and the actual flux cut at a density of B Wb per sq. m = $B \times \ell \times d$, hence

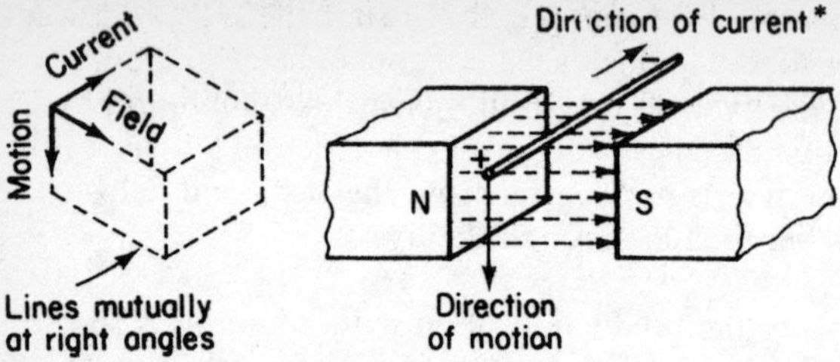

(i) Conductor moving through a magnetic field.

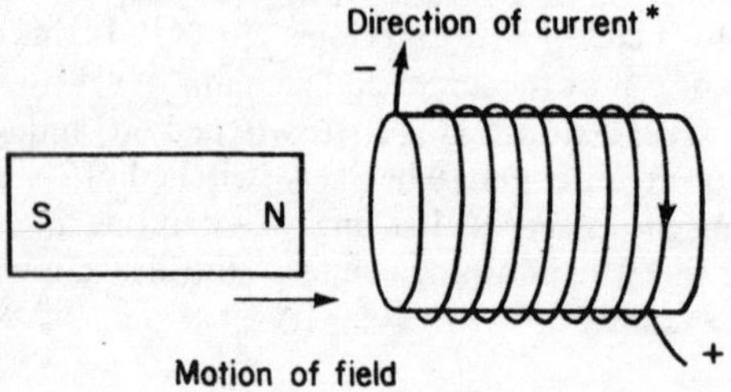

(ii) Production of e.m.f. in a coil of wire using a bar magnet.

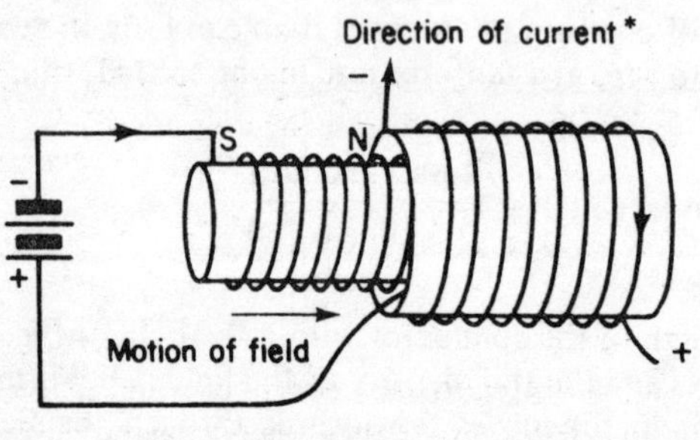

(iii) Production of e.m.f. in a coil of wire using an energized coil.

* Electron flow
(when external circuit connected).

Fig. 5.7 Electromagnetic induction.

$$\text{e.m.f. (e)} = \frac{B\ell d}{t} \text{ volts}$$

At constant velocity (uniform rate of cutting flux)

$$\text{Velocity, } v = \frac{d}{t} \text{ metres per sec}$$

(velocity is speed or rate of motion in a given direction and is equal to the distance travelled in a certain time).

$\therefore$ induced e.m.f. $e = B\ell v$ volts

EXAMPLE:
A straight conductor of length 10 cm is moved at a rate of 100 m/s at right angles to a magnetic field of 0.5 T (Wb/m^2). What e.m.f. is induced?

$B = 0.5$ T $\quad \ell = 10 \times 10^{-2}$ m $\quad v = 100$ m/s

$$\text{e.m.f. induced} = B\ell v \text{ volts} = 0.5 \times 10 \times 10^{-2} \times 100$$

$$= \underline{5 \text{ volts}}$$

Regarding Fig. 5.7(ii), a flux change of 1 Wb/s induces an e.m.f. of 1 volt into each turn of the coil, these e.m.f.s are additive, therefore

$$e = \text{No. of turns} \times \text{rate of flux change}$$

EXAMPLE:
A bar magnet is thrust into the centre of a coil of 1000 turns so that the flux threading the coil increases at a uniform rate of 0.5 Wb per second. The total circuit resistance (external circuit plus coil) is 100 ohms. Calculate the momentary current which flows.

$$\text{Induced e.m.f.} = 1000 \text{ turns} \times 0.5 \text{ Wb/s} = 500 \text{ V}$$

From Ohm's Law

$$I = \frac{500}{100} = \underline{5 \text{ amperes}}$$

Sir John Ambrose Fleming (an English electrical engineer) gave us a convenient rule for finding the direction of the e.m.f. induced under these conditions. Here we must not lose sight of the fact that we are studying *generators*, that is devices which produce e.m.f.s and the e.m.f. of the generator in action, be it only a piece of wire moving in a magnetic field, will have + and – polarities. Current will flow in the external circuit from the terminal with an excess of electrons (–) to the one with a deficiency (+) but *within* the generator, as in the case of a battery, the current must flow from the + terminal to the – terminal. Fig. 5.7 is labelled to illustrate the point. Also in (i) it can be seen that three directions are considered at once, the flux travels horizontally (left to right), the wire vertically (downwards) and the current diagrammatically, travels into the paper. We say that these directions are all mutually (i.e. with reference to each other) at right angles. Pictorially, they may be represented by a particular corner of a box as also shown in Fig. 5.7(i). From Fleming's suggestion we use a Left-Hand Rule by extending thumb, forefinger and middle finger to point in the three directions mutually at right angles as with the corner of the box. [Note that Fleming's original rules used the opposite hands because they were associated with "conventional" current (Sect. 1.2.4).]

If now the left hand is turned so that the

> Fore-Finger points in the direction of the Flux or Field
>
> the thuMb points in the direction of the Motion, then
>
> the mIddle fInger points in the direction of the Current (I), or if preferred,
>
> the Middle Finger points in the direction of the e.M.F.

We shall find a convenient way of remembering which hand to use later in the chapter.

5.3.1 Inductance

When current flows in a coil of wire as, for example, in Fig. 5.3(i), the magnetic flux it creates is proportional to that current (Section 5.2.3). Therefore, whenever the current changes, the magnetic flux changes proportionally. But as we have just seen, when flux linking a coil changes, an e.m.f. is induced in the coil, so in effect the original current is, through its magnetic field, producing an e.m.f. This is known as *self-induction* because it all happens within the same coil and the coil is said to possess *self-inductance*, or more commonly just *inductance.*

A further law applies here and fortunately it is an easy one to remember. Heinrich Lenz (a Russian physicist) stated that the direction of an induced e.m.f. is such that its own electromagnetic effect is to oppose the one producing it. This is known as Lenz's Law. Stated in a slightly different way it says that every induced current delays the change of flux density which is producing it.

Because the induced e.m.f. gives rise to a current in opposition to the original one, it is known as the *counter or back e.m.f.*

Looking at the whole process of self-induction in stages:

(i) a voltage is applied and current starts to build up in the coil;
(ii) therefore a magnetic flux grows;
(iii) because the magnetic flux is changing and cuts the same turns of the coil, an e.m.f. is induced in those turns;
(iv) Lenz's Law shows that the e.m.f. will cause a current to flow in opposition to the applied current;
(v) hence the applied current and magnetic flux will grow more slowly than they would under non-inductive conditions.

Equally, when the applied current is switched off, the collapsing magnetic field sets up a back e.m.f. which causes a current to flow in such a direction as to tend to maintain the flux.

The unit of self-inductance is the *Henry* (symbol H, after Joseph Henry, a U.S. physicist). Inductance itself is denoted by the symbol L. A coil or circuit has an inductance of one henry if an e.m.f. of one volt is induced when the current changes uniformly at the rate of one ampere per second. We might conveniently sum up inductance as being that electrical characteristic which creates opposition to a change of current.

We can calculate any other rate of variation of current other than one ampere per second by dividing the actual change by the time it takes, for example, a current rising from I_1 to I_2 amperes in a time of t seconds is at a rate of $(I_2 - I_1)/t$ amperes/second (A/s).

Now if 1 V is induced in 1 H when the current changes at 1 A/s, we can write a general formula

$$L = \frac{e}{\text{rate of change of current}}$$

where e is the induced e.m.f.

$$\therefore \quad e = L \times \frac{I_2 - I_1}{t} \text{ volts}$$

EXAMPLE:
An electromagnetic relay coil has an inductance of 3 henrys. The full current is 2 amps. If this is switched off and falls to zero in 50 ms, what is the average value of the e.m.f. induced?

Rate of change of current =

$$2 \text{ A in } 50 \text{ ms} = \frac{2}{50 \times 10^{-3}} = \frac{2 \times 10^2}{5} \text{ A/s}$$

Then, $e = L \times \text{rate of change of current}$

$$= 3 \times \frac{2 \times 10^2}{5} = \underline{120\ \text{V}}$$

from which we see that quite high e.m.f.s can be induced when the flux changes rapidly, often much higher than the one which originally set up the magnetic field. Many car owners know this only too well if they have had the shocking experience of touching a bare plug lead when that particular plug is firing. A magnetic flux is collapsing rapidly through a coil of thousands of turns of fine wire, thus producing a very high induced e.m.f. (around 20,000 V). Contact with this is not fatal (Section 3.6.2) since the *pulse* of voltage is of very short duration and also because the internal resistance (Section 3.3) of the source (the coil) is high, as soon as even a tiny current starts to flow, voltage is lost across it. It follows that even though the induced e.m.f. is high, very little current can flow in an external circuit.

Calculation of inductance:
This is more of an art than a science for several reasons, the main one being that the flux in a coil does not follow a neat path except when confined within a high permeability material. Even then the difficulty arises that variation in flux is not directly proportional to the variation in current because the relative permeability, μ_r, falls as flux density increases. Nevertheless it is instructive to obtain a formula for an iron-cored coil with a closed magnetic circuit (e.g. Fig. 5.4) on the basis that μ_r remains constant. From such a formula the effect of each parameter which contributes towards the final value of inductance will be seen.

Suppose a coil of N turns has a uniform rise of current within it from 0 to I amps in t secs.

Induced e.m.f. e = L x rate of change of current

$$= L \times \frac{I - 0}{t} = \frac{LI}{t} \text{ volts}$$

and from the definition of the weber,

$$e = \frac{\text{total flux cut}}{\text{time}} \text{ for 1 turn}$$

$$\therefore \quad e = \frac{\Phi}{t} \text{ for 1 turn}$$

$$= \frac{N\Phi}{t} \text{ for N turns}$$

Equating the two expressions for e,

$$\frac{LI}{t} = \frac{N\Phi}{t}$$

$$\therefore \quad L = \frac{N\Phi}{I}$$

But $$\Phi = \frac{F}{S} = \frac{4\pi NI\mu_r a}{10^7 \ell}$$

as already determined, where μ_r is the relative permeability of the core, ℓ and a are its dimensions.

∴ substituting for Φ,

$$L = \frac{N}{I} \times \frac{4\pi NI\mu_r a}{10^7 \ell} = \underline{\frac{4\pi N^2 \mu_r a}{10^7 \ell}}$$

thus the inductance increases with the *square* of the number of turns, the cross-sectional area and permeability of the core, but decreases with its length.

EXAMPLE:
A straight air-cored solenoid (cylindrical coil of wire) has 5000 turns. It has a diameter of 5 cm and is 50 cm long. What is the approximate inductance?

$$N = 5000 \qquad a = \frac{\pi}{4}(5 \times 10^{-2})^2 \text{ sq. m}$$

$$\ell = 0.5 \text{ m} \qquad \mu_r = 1 \qquad L = ?$$

Then,

$$L = \frac{4\pi N^2 \mu_r a}{10^7 \ell} = \frac{4\pi \times 5000^2 \times 1 \times \pi \times 25 \times 10^{-4}}{10^7 \times 4 \times 0.5}$$

$$= \frac{125\pi^2}{10^4} \text{ H} = \frac{125\pi^2}{10} \text{ mH} = \underline{123.37 \text{ mH}}$$

EXAMPLE:
A steel ring has a winding of 1000 turns on it. The dimensions of the ring are, mean radius 10 cm and cross-sectional area 2.5 sq. cm (see Fig. 5.4). If the relative permeability can be taken as 2000, what is the inductance of the coil?

$$N = 1000 \qquad a = 2.5 \times 10^{-4} \text{ sq. m}$$

$$\ell = 20\pi \times 10^{-2} \text{ m} \qquad \mu_r = 2000 \qquad L = ?$$

Then,

$$L = \frac{4\pi N^2 \mu_r a}{10^7 \ell} = \frac{4\pi \times 1000^2 \times 2000 \times 2.5 \times 10^{-4}}{10^7 \times 20\pi \times 10^{-2}} \text{ H}$$

$$= \frac{4 \times 5}{20} = \underline{1.0 \text{ H}}$$

5.3.2 *Mutual inductance*

Just as a changing magnetic field set up by current in a coil induces an e.m.f. in that coil so equally an e.m.f. will be induced in any other coil within the sphere of influence of the field. This is called *mutual inductance* and is represented diagrammatically as in Fig. 5.8(i) where a current in Coil 1 sets up a magnetic flux linking not only itself but also the separate Coil 2. An e.m.f. is induced in Coil 2 when the current changes in 1, but only when the current varies, not when it has reached a steady value. The figure shows what is most frequently called a "transformer", that is, two or more coils sharing a common magnetic circuit. Transformers are more easily recognized when they have a core and one of the common uses of this type is in electronic equipment for

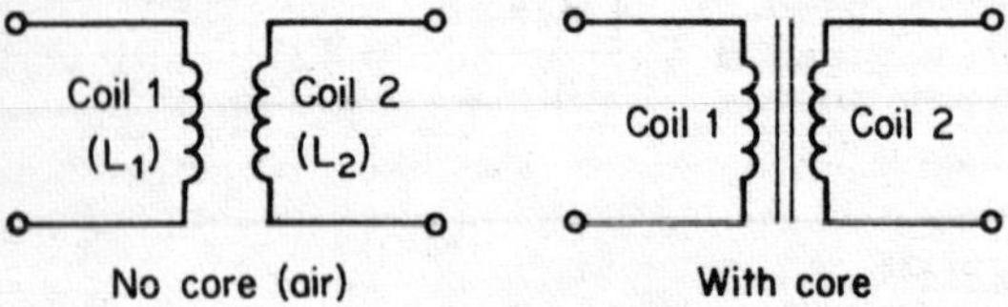

(i) General symbols.

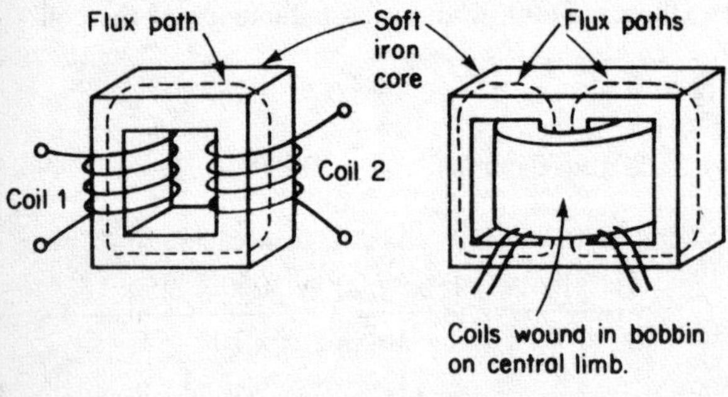

(ii) Practical arrangements.

Fig. 5.8 Transformers.

transforming the electricity supply into a form suitable for transistor circuits. In practice the transformer with a core may appear as in Fig. 5.8(ii) where in either case, one coil, by carrying a varying current can set up a changing magnetic flux around the core and therefore embrace the second coil. In the second arrangement, which is more frequently used, the magnetic flux has two paths in parallel.

Obviously the nearer the two coils are to each other, the closer or tighter the *coupling* and the greater will be the effect of one coil on the other. This leads to a definition of mutual inductance. Two coils have a mutual inductance of one henry if an e.m.f. of one volt is induced in one coil when the current changes uniformly at the rate of one ampere per second in the other.

How tightly the coils are coupled together (they may be centimetres apart or conversely one wound on top of the other) is expressed by a factor known as the *coefficient of coupling* and denoted by k. If we imagine two coils 1 and 2 coupled as in Fig. 5.8(i), then if all the flux generated by Coil 1 cuts Coil 2, $k = 1$, if the coils are moved apart until only half the flux cuts Coil 2, $k = 0.5$. In practice, k can never quite reach the value of 1 for there is always a little flux which is ineffective, nevertheless in transformers such as depicted in the second arrangement of Fig. 5.8(ii), k can be as high as 0.98 to 0.99.

Since both the flux (Φ) and the e.m.f. induced (e) are proportional to the number of turns, a certain value of Φ or e is multiplied by N when the number of turns is so increased. Considering both coils to have one turn each and to be completely coupled ($k = 1$), with a current change in Coil 1 giving rise to an e.m.f. e in Coil 2, then if the number of turns on Coil 1 is increased to N_1 and in addition the number of turns on Coil 2 is increased to N_2, the induced e.m.f. becomes $N_1 N_2 e$, that is, the mutual inductance (M henrys) has increased by a factor $N_1 N_2$, i.e. $M \propto N_1 N_2$.

Now from the previous section, the inductance of Coil 1, $L_1 \propto N_1^2$ and similarly $L_2 \propto N_2^2$, hence

$$L_1 \times L_2 \propto N_1^2 \times N_2^2$$

or, taking square roots,

$$\sqrt{L_1 L_2} \propto N_1 N_2,$$

i.e. $M \propto \sqrt{L_1 L_2}$

and if the coupling factor k is brought into effect,

$$\underline{M = k\sqrt{L_1 L_2}}$$

The main use of a transformer is to transfer power from one winding called the *primary* to the other, known as the *secondary*. It can thus couple two circuits magnetically, with no direct metallic connexion between them. This is a very useful feature used in mains-driven transistorized equipment to isolate the equipment which probably works at only a few volts, from the mains supply which is frequently as high as 240 volts. The other main function is to transform voltage from one value to another as above, say from 110 or 240 V to 12 V. But note that these are alternating voltages (Section 3.7) which we have yet to consider (Book 2), a direct voltage applied to one winding of a transformer will have no effect on the second winding once the few milliseconds have elapsed for the current to reach its steady state.

If a transformer has Np turns on its primary and Ns turns on its secondary winding, it is said to have a *turns-ratio* of Ns/Np and since the voltage induced in each turn of the secondary winding is the same as the voltage which is self-induced in each turn of the primary (all turns are sharing equally the effect of the same, varying magnetic field), then the ratio of primary and secondary voltages is the same as the ratio of primary to secondary turns, i.e.

$$\frac{Vs}{Vp} = \frac{Ns}{Np}$$

where Vp and Vs are the primary and secondary voltages and Np and Ns their numbers of turns.

A transformer is labelled *step-up* when Vs is greater than Vp and *step-down* when Vs is less than Vp.

EXAMPLE:
A transformer is required for operation of 10 V transistorized equipment from a 200 V mains supply. If there are 2000 turns on the primary winding which is connected to the mains, how many turns will be required on the secondary winding?

$$Vp = 200 \qquad Vs = 10 \qquad Np = 2000 \qquad Ns = ?$$

Then,

$$\frac{Vs}{Vp} = \frac{Ns}{Np}$$

$$\therefore \quad \frac{10}{200} = \frac{Ns}{2000} \quad \text{(cross-multiply)}$$

$$\therefore \quad 10 \times 2000 = 200 \times Ns$$

$$\therefore \quad Ns = \frac{10 \times 2000}{200} = \underline{100 \text{ turns}}$$

The transformer is therefore a step-down type with a turns-ratio

$$\frac{Ns}{Np} = \frac{100}{2000}, \text{ i.e. } 0.05{:}1.$$

In practice turns-ratios are sometimes rearranged so that the ratio is greater than 1, in this case the ratio would be quoted as (Np/Ns) = (2000/100), i.e. 20:1. Generally the use of the transformer will make it obvious which winding has the greater number of turns.

5.4 Practical Inductors and Transformers

A straight length of wire has inductance, when coiled the inductance increases and we then recognize it as an inductor, with two or more coils magnetically coupled together it becomes a transformer as already shown in Fig. 5.8(ii). Two of the many other types are sketched in Fig. 5.9. At (i) is a simple radio inductor which is variable, the iron-dust *slug* is screwed down into the centre of the coil to raise the inductance by decreasing the reluctance of the magnetic circuit. Such coils have inductance values from microhenrys to a few millihenrys.

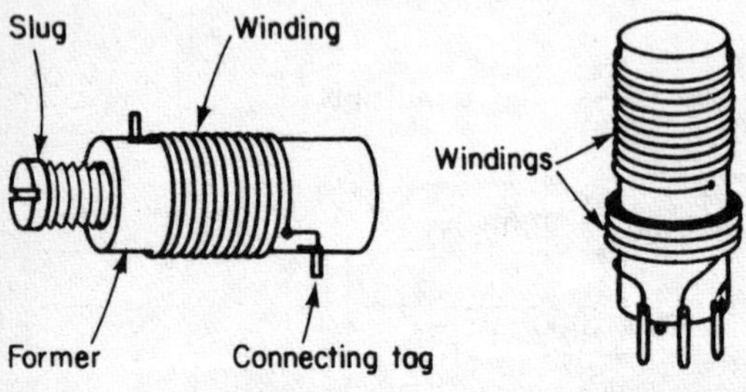

(i) Variable radio inductor. (ii) Radio transformer.

Fig. 5.9 Typical variable inductor and radio transformer.

A type of radio receiver transformer is illustrated at (ii). This has an air-core and a low coupling factor.

Frequently inductors and transformers are *encapsulated* in an epoxy-resin for protection against moisture and shock. When this happens the component is no longer recognizable as it will appear merely as a rectangular block of hard resin with connecting wires or tags protruding.

5.5 Inductors in Series and Parallel

When inductors are interconnected there could be a complication due to the magnetic flux of one inductor affecting a second, it will be assumed that this does not occur.

It will be found that inductors in series and parallel follow the same form of relationship as for resistors.

Series:
If two inductors having inductances L_1 and L_2 are connected in series as in Fig. 5.10(i), the rate of change of current $(I_2 - I_1)/t$ is the same in both.

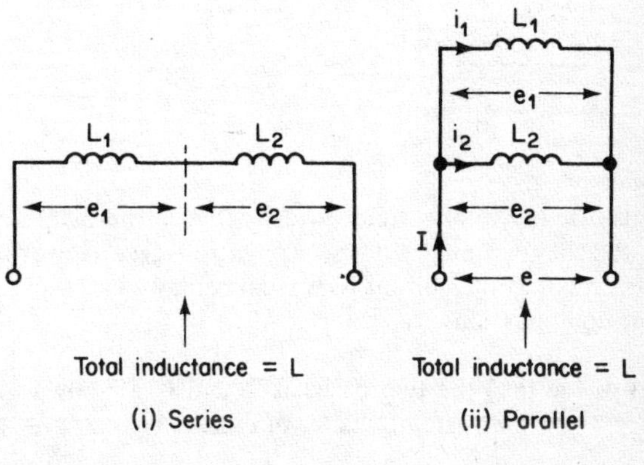

Fig. 5.10 Inductors connected in series and parallel.

Hence, since

$$e = L \times \left(\frac{I_2 - I_1}{t}\right)$$

then,

$$e_1 = L_1 \left(\frac{I_2 - I_1}{t}\right)$$

and

$$e_2 = L_2 \left(\frac{I_2 - I_1}{t}\right)$$

and total e.m.f.

$$e = e_1 + e_2 = (L_1 + L_2)\left(\frac{I_2 - I_1}{t}\right)$$

∴ total inductance $L = L_1 + L_2$ and similarly it can be shown that

$$\underline{L = L_1 + L_2 + L_3 + \ldots}$$

Parallel:
In this case the total current divides between the two inductors and the reasoning is not so straightforward. However, if we assume that the current rises from zero, we can avoid some of the complications.

Let I divide into the two paths as i_1 and i_2 as shown in Fig. 5.10(ii), then rate of change of current in

$$L_1 = \frac{i_1 - 0}{t} = \frac{i_1}{t}$$

and in L_2, $\frac{i_2}{t}$

$$e_1 = L_1 \frac{i_1}{t} \qquad \therefore \frac{i_1}{t} = \frac{e_1}{L_1}$$

and $$e_2 = L_2 \frac{i_2}{t} \qquad \therefore \frac{i_2}{t} = \frac{e_2}{L_2}$$

also $$e = L\frac{I}{t} \qquad \therefore \frac{I}{t} = \frac{e}{L}$$

But $e_1 = e_2 = e$ (all connected together)

$$\therefore \quad \frac{i_1}{t} = \frac{1}{L_1} \qquad \frac{i_2}{t} = \frac{1}{L_2} \qquad \frac{I}{t} = \frac{1}{L}$$

$$\therefore \quad i_1 = \frac{1}{L_1} \qquad i_2 = \frac{1}{L_2} \qquad I = \frac{1}{L}$$

and since $i_1 + i_2 = I$

then

$$\frac{1}{L} = \frac{1}{L_1} + \frac{1}{L_2}$$

and similarly

$$\frac{1}{L} = \frac{1}{L_1} + \frac{1}{L_2} + \frac{1}{L_3} + \ldots$$

5.6 The Magnetic Field at Work

When a conductor which is part of a circuit moves in a magnetic field, a current flows. The converse is also true for when a current flows in a conductor which lies across a magnetic field the conductor will move through interaction between its own field and the external one, provided that it is free to do so. Thus the magnetic field does work, the basic principle of the electric motor.

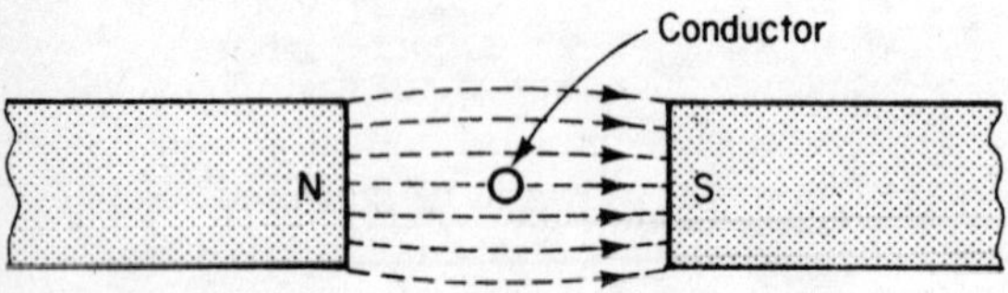

(i) Conductor situated in magnetic field.

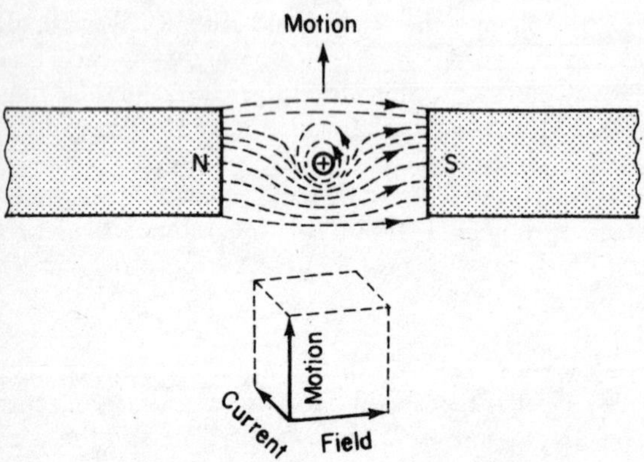

(ii) Combined fields when current flows.

Fig. 5.11 Electromagnetic force on a conductor.

Consider a conductor situated between a pair of N and S poles as shown in Fig. 5.11(i). When no current flows, the field is undisturbed but when the conductor carries a current, its own circular flux reacts with that of the magnet, combining when the two flux directions coincide but cancelling to a certain degree when they are opposite. A line drawing representing the combined field is shown at (ii) and in this particular example it is evident that there is a crowding of flux lines below the conductor with a depletion above. A force is exerted on the conductor trying to move it from the stronger part of the field to the weaker, in effect as though the lines were elastic and attempting to straighten.

A Right-Hand Rule is used in this, the motor, case with the thumb and two fingers extended mutually at right angles as before (Section 5.3) and with the same indications – also shown by the corner of the box in Fig. 5.11(ii).

The previous section dealt with the Generator Principle for which the Left-Hand Rule is appropriate, here we have the Motor Principle and the Right-Hand Rule. The difficulty for most is not so much in remembering what each finger stands for (thumb, motion; forefinger, flux; middle finger, current) but in remembering which hand, left or right. It may be useful to recall that quality reserved exclusively for the ladies, "GLAMOUR". Drop the vowels and we are left with four consonants, GLMR for Generator, Left; Motors, Right. Remember this is for electron current flow, not "conventional" current.

The most obvious example of the motor principle is, of course, shown by electric motors, in one form wires carrying current are fixed within a central rotating *armature*. Around this is a radial magnetic field forcing the wires to move according to the Right-Hand Rule, that is, the armature rotates. An external switching arrangement fixed to the shaft of the armature and known as a *commutator* reverses the direction of the current in the armature wires as they move from the influences of a N to a S pole and vice versa so that the force always turns the armature in the same direction.

This is more a subject for the power engineer, of greater interest in electronics is the application of the motor principle for current measurement.

5.6.1 Moving-coil instruments

Bearing in mind that the measurement of any electrical characteristic is dealing with the unseen, it is not unreasonable to expect difficulty. An example is in the measurement of capacitance and inductance where an uncomplicated meter

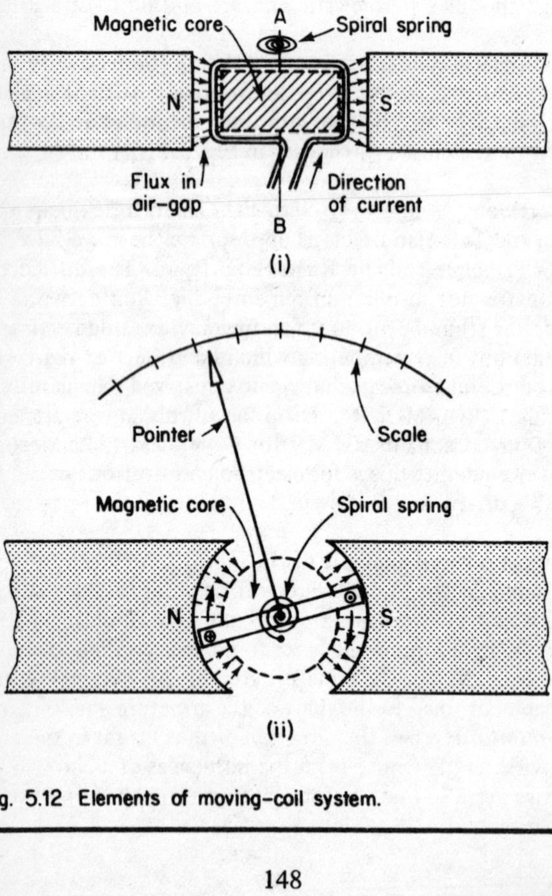

Fig. 5.12 Elements of moving-coil system.

or instrument giving a direct reading on a dial is not easily designed. However for measuring current a simple meter is possible, based on the motor principle and indicating the current magnitude by a pointer moving over a scale.

The *moving-coil* meter has been used almost exclusively in electrical work for a long time, to some extent being superseded by the more recently developed digital instrument in which the value is displayed numerically instead of being read off from a pointer moving over a dial. It is not likely to replace the earlier type completely because each has its own advantages.

A side view of the simplest form of moving-coil system in which the coil consists of only one turn is shown in Fig. 5.12(i). N and S are the poles of a permanent magnet providing a magnetic field in which the coil is free to turn on the spindle AB. A plan of the arrangement is shown at (ii) with the pointer and scale added. With current flowing round the coil in the direction shown, the Right-Hand Rule shows that the coil will turn as indicated by the arrows, causing the pointer to move up the scale against the restoring torque of the spiral spring. When the current ceases, the spring returns the pointer to the rest (zero) position. The practical coil has many turns and it rotates around a magnetic core which serves to reduce the magnetic circuit reluctance to a minimum (Section 5.2.2) and to keep the lines of flux at right angles to the coil. In this instrument the deflexion of the needle is proportional to the current in the coil. Movements requiring only 50 μA for a full-scale deflexion are quite common.

Appendix 1

ABBREVIATIONS

Most professions have their own particular abbreviations and electronics with its use of mathematics is no exception. By their use many words and phrases can be saved.

Literary

Fig.	Figure
Chap.	Chapter
Sect.	Section (this book has 5 chapters, each divided into sections)
approx.	approximately
i.e.	Latin, id est – that is to say
e.g.	Latin, exempli gratia – for, by the way of, example
l.h./r.h.	left hand/right hand
l.h.s./r.h.s.	left-hand side/right-hand side
vice versa	Latin – by transposing the main items in the statement just made.

Mathematical

$=$	is equal to
$\equiv$	is equivalent to
$\approx$	is approximately equal to
$\neq$	is not equal to
$\therefore$	therefore
$\rightarrow$	tends to or approaches
$>$	is greater than
$<$	is less than
$\sqrt{}$	square root of
$\sqrt[n]{}$	nth root of
$\angle$	angle
rt $\angle$	right angle
$\propto$	varies as

Electronics

+, +ve	positive
–, –ve	negative
e.m.f.	electromotive force
p.d.	potential difference

Appendix 2

DECIMALS

Decimal notation, that is, counting in tens, is the basis of most arithmetical systems. An important feature of any decimal number is that each "place" in the number indicates a different multiple of ten:

No. of hundreds	No. of tens	No. of units		No. of tenths	No. of hundredths
8	6	4	.	9	2

and the decimal point always follows the units number and so separates the whole number part from the fractional (less than one) part.

Now if we were to multiply the above number by 10, this would be equivalent to changing the units to tens, tens to hundreds, hundredths to tenths and so on:

No. of thousands	No. of hundreds	No. of tens	No. of units		No. of tenths	No. of hundredths
8	6	4	9	.	2	0

and note that all we have actually done is to shift the decimal point one place to the right. It follows that to multiply a number by 100 the decimal point is moved two places to the right, in fact the number of places the decimal point is moved is the same as there are noughts in the multiplier, a very convenient arrangement.

Equally to divide a number by 10 or a multiple of 10, the decimal point is shifted to the left according to the number of noughts in the divisor (number by which another number is divided). Therefore:

(i) $864.92 \times 10 = 8649.2$
(ii) $864.92 \times 100 = 86492$
(iii) $864.92 \div 100 = 8.6492$

and we see that in (ii) the decimal point has disappeared because we do not normally write one when there is no fractional part. However, it must be remembered that in fact:

86492 can equally be written as 86492.0000 (as many noughts added as we may need, they add up to nothing so do not change the value of the number)

so that $864.92 \times 10{,}000 = 864.9200 \times 10{,}000 = 8649200$

The same happens in reverse when dividing, for 864.92 can be written as 00864.92 again with as many noughts preceding the number as required so that

$$864.92 \div 10{,}000 = 0.086492$$

It will quickly be realised from the main text that in electronics we often have to handle very large or very small numbers and frequently many noughts are involved. Not only is this inconvenient but writing down so many each time allows errors to creep in, thus we use the standard mathematical shorthand method which at first looks difficult but with use will be found to be simplicity itself. Now:

10×10 we call "10 squared", write it as 10^2 and it is equal to 100.

$10 \times 10 \times 10$ we call "10 cubed", write it as 10^3 and it is equal to 1,000.

$10 \times 10 \times 10 \times 10$ we call "10 to the fourth", write it as 10^4 and it is equal to 10,000.

When we write the numbers, 10 is called the *base* and the little raised figure is called the *exponent* or *index.*

For numbers greater than one, the number of noughts in the fully written figure is the same as the exponent in the short-hand form. Not much saved by writing 10,000 as 10^4 but in Section 1.2, for example, it is found that there are about 10^{23} electrons free in a certain piece of wire. To write 10^{23} as 1 followed by 23 noughts is wasteful when the simple form 10^{23} ("ten to the twenty-third power" or simply "ten to the twenty-third") means the same. For numbers less than 1 the number of noughts after the decimal point in the fully written figure is one less than the number in the exponent in the shorthand form. Hence:

$$
\begin{aligned}
1000 &= 10^3 \\
100 &= 10^2 \\
10 &= 10^1 \\
1 &= 10^0 \\
\tfrac{1}{10} \text{ i.e. } 0.1 &= 10^{-1} \\
\tfrac{1}{100} \text{ i.e. } 0.01 &= 10^{-2} \\
\tfrac{1}{1000} \text{ i.e. } 0.001 &= 10^{-3}
\end{aligned}
$$

Going down the left-hand side, each line has been divided by 10 to obtain the next whereupon the exponent has been reduced by 1 so that from line 1 to line 2:

$$\frac{1000}{10} = 100\,, \text{ i.e. } 10^{3-1} = 10^2$$

and conversely, going up the left-hand side, each line is multi-plied by 10 causing 1 to be added to the exponent, for example:

$$0.01 \times 10 = 0.1, \text{i.e. } 10^{-2+1} = 10^{-1}$$

Thus dividing by 10 decreases the exponent by 1, dividing by 100 decreases it by 2 and so on, according to the number of noughts in the divisor. Multiplying by 10, 100, etc. increases the exponent in the same way and in general, by letting a and b represent any two exponents:

$$10^a \times 10^b = 10^{a+b} \quad \text{and} \quad 10^a \div 10^b = 10^{a-b}$$

and note from the second that dividing by 10^b is the same as multiplying by 10^{-b} so we could have written

$$\frac{10^a}{10^b} = 10^a \times 10^{-b} = 10^{a-b}$$

i.e. shifting a multiple of ten from numerator to denominator or denominator to numerator changes the sign of the exponent,

$$\frac{1}{10^b} = 10^{-b} \quad \text{and} \quad \frac{1}{10^{-b}} = 10^b$$

Numbers not commencing with 1 can also be shortened according to the above decimal and exponent rules, for example, 70,000 can be expressed as

$7 \times 10{,}000$, i.e. 7×10^4

or even

70×10^3

A little more complicated is 6,250,000, which is equal to

6.25×10^6 or 625×10^4

[remembering that if we have an expression showing two quantities multiplied together, the value of the expression is unchanged if we divide one of the quantities by another number and multiply the other quantity by the same number, e.g. 6.25×10^6 – multiply 6.25×100, divide 10^6 by 100, gives

$$(6.25 \times 100) \times \frac{10^6}{100}, \text{ i.e. } 625 \times (10^6 \times 10^{-2})$$

or 625×10^4]

EXAMPLE:
If one coulomb of electricity is carried by 6.25×10^{18} electrons, how much is carried by each?

Quantity carried by one electron

$$= \frac{1}{6.25 \times 10^{18}} = \frac{1}{6.25} \times \frac{1}{10^{18}}$$

$$= 0.16 \times 10^{0-18} = 0.16 \times 10^{-18} \text{ coulombs, or}$$

$$\underline{1.6 \times 10^{-19} \text{ coulombs}}$$

(6.25×10^6 and 1.6×10^{-19}, i.e. where there is only one digit to the left of the decimal point, express the number in *scientific notation*.)

EXAMPLE:
A strip of copper contains 8×10^{24} atoms. If one in every hundred atoms gives up an electron, how many free electrons will there be?

We find how many hundreds (10^2) of atoms there are and each of these gives rise to one free electron. Therefore:

$$\text{No. of free electrons} = (8 \times 10^{24}) \div 100$$

$$= 8 \times 10^{24} \times 10^{-2} = 8 \times 10^{24-2} = \underline{8 \times 10^{22}}$$

EXAMPLE:
The charge on an electron is 1.6×10^{-19} coulombs. What is the total charge on two million electrons?

$$1 \text{ million} = 1{,}000{,}000 = 10^6$$

$$2 \text{ million} = 2 \times 10^6$$

$$\therefore \quad \text{Total charge} = 1.6 \times 10^{-19} \times 2 \times 10^6$$

$$= 3.2 \times 10^{-19+6} = \underline{3.2 \times 10^{-13} \text{ coulombs}}$$

Appendix 3

LOGARITHMS

In Appendix 2 a general rule regarding exponents is developed:

$$10^a \times 10^b = 10^{a+b}$$

and

$$10^a \div 10^b = 10^{a-b}$$

On the l.h.s. of each expression is multiplication or division equalled in both cases on the r.h.s. by a term involving addition or subtraction. With large numbers or numbers with several decimal places, multiplication or division is much more difficult to carry out than addition or subtraction, thus a system which converts the former to the latter has much to recommend it. Tables of logarithms do this with an accuracy more than sufficient for most electronic work.

In the formulae above, 10 is known as the *base* and a and b are *exponents* or *indices*. Any other value may be used for the base but 10 has many advantages and is in general use; logarithms to the base of 10 are known as *common logarithms*. Now we already know from Appendix 2 that, for example, 1, 10, 100, 1000 . . . are also written 10^0, 10^1, 10^2, 10^3 . . . , the common logarithm is simply the exponent in each case so that:

the logarithm of	1	is 0
” ” ”	10	is 1
” ” ”	100	is 2
” ” ”	1000	is 3
and of	0.1	is −1
” ”	0.01	is −2

What we have yet to determine are the values between 1 and 10, 10 and 100, etc., and these are given to us in common logarithm tables (the term "common" is usually dropped). These give the exponent (or logarithm) for all values from 1 to 9.999, for example

> the logarithm of 2 is 0.3010 (shown against 20) meaning that $10^{0.3010} = 2$, and
>
> the logarithm of 9 is 0.9542 (shown against 90) meaning that $10^{0.9542} = 9$

and using the four-figure tables to the full, the logarithm of, say, 5.647 is 0.7518, meaning that

$$10^{0.7518} = 5.647.$$

This is read horizontally from 56 in the left-hand column, and in the main column under 4, giving a logarithm for 5640 of 0.7513. To this is added the figure given under 7 in the right-hand columns which in this case is 5, hence the logarithm of 5.647 is read as 0.7513 + 5, i.e. 0.7518.

Now, if we need the logarithm of say 56.47, which is ten times 5.647, recalling that a logarithm is simply the exponent and that multiplying by 10 adds 1 to the exponent, then we automatically have the logarithm for 56.47 as

$$0.7518 + 1 = 1.7518$$

A useful way of ensuring that the number before the decimal point is correct is to insert the answers within a table of values which can easily be recalled as follows:

$$\begin{aligned} \text{log. } 1 &= 0 \\ \text{log. } 5.647 &= 0.7518 \\ \text{log. } 10 &= 1 \\ \text{log. } 56.47 &= 1.7518 \\ \text{log. } 100 &= 2 \end{aligned}$$

thus because 5.647 is between 1 and 10, its logarithm must be greater than 0 but less than 1. Similarly log. 56.47 must be greater than 1 but less than 2. For the reader who needs practice, it is suggested that logarithms of the following should be found, the answers are given in brackets.

EXAMPLE:
Find the (common) logarithms of the following:

8042 (3.9054)	26 (1.4150)	26.2 (1.4183)
562.8 (2.7503)	71892 (4.8566)	

(Note – in the last case the logarithm of a 5-figure number is required, ordinary logarithm tables are 4-figure (7-figure can also be obtained but we shall seldom need the greater accuracy they give). Therefore in this particular case, the units figure of 2 must be ignored, that is, the logarithm of 71890 is obtained.)

Straightforward so far but we now come to the part where many students have difficulty, which is in finding and manipulating the logarithms of numbers less than 1. The rules given show that dividing by 10, 100, . . . , reduces the logarithm by 1, 2, . . . , and taking one of the examples above:

log. 26 = 1.4150
log. 2.6 = 1.4150 – 1 = 0.4150
log. 0.26 = 1.4150 – 2 (or equally 0.4150 – 1)

and this would be equal to –0.5850. Quite correct, but showing that a new set of tables would be needed for logarithms of numbers less than 1 because of the negative logarithm. The extra set of tables is avoided by considering these logarithms as being made up of two separate parts, that before the decimal point which incidently is known as the *characteristic* and for any logarithm can be positive, negative or zero, and the four figures after the decimal point known as the *mantissa* which is always positive. So that there is no confusion with a negative characteristic, the minus sign is placed above the number, not preceding it, for this would make the whole logarithm negative. So we write log. 0.26 not as –0.5850 but $\bar{1}.4150$ remembering that this is actually equivalent to $-1 + 0.4150$, and again

log. 0.026	=	$\bar{2}.4150$
log. 0.0026	=	$\bar{3}.4150$, etc.

Although the table of multiples of ten method above of checking is still valid we can also produce a table for obtaining the value of the characteristic directly as in Table A3.1(i).

The mantissa is therefore obtained directly from the standard logarithm table so this and Table A3.1(i) give the means of converting any number to its logarithm.

EXAMPLE:
Find the logarithms of the following:

(i) 3 (0.4771)
(ii) 300 (2.4771)
(iii) 0.3 ($\bar{1}.4771$)
(iv) 0.0003 ($\bar{4}.4771$)

A3.1 Multiplication

To multiply two or more numbers together, add their logarithms to obtain the logarithm of the product. The number which corresponds to this logarithm (called the *anti-logarithm*) is the answer.

Anti-logarithms are usually supplied as a second table to logarithms, they speed the work although are not essential because the logarithm table itself can be used in the inverse manner, that is by searching through the logarithms to find the number corresponding.

TABLE A3.1 Rules Concerning Characteristics of Logarithms

(i) Conversion from a number to its logarithm

Number	*Characteristic*
Greater than 1	Positive. One less than the number of figures to the left of the decimal point.
Less than 1	Negative. One more than the number of noughts to the right of the decimal point.

(ii) Conversion from a logarithm to its number

Characteristic	*Number*
Positive or Zero	Number of figures to left of decimal point equal to (characteristic + 1)
Negative	Number of noughts to right of decimal point equal to (characteristic – 1)

A check can always be made on the method of calculating a known sum by logarithms, e.g. 9 x 4

log. 9	= 0.9542
log. 4	= 0.6021
log. (9 x 4)	= 1.5563

antilog 0.5563 = 3599

Table A3.1(ii) helps in conversion of logarithms back to their corresponding numbers and shows that a positive characteristic of 1 requires 2 figures to the left of the decimal point, i.e. antilog 1.5563 = 35.99.

The correct answer is, of course, 36 which while confirming the method also shows that 4-figure logarithms do allow some inaccuracy to creep in (with 7-figure logarithms the answer is 36.00020, almost correct).

The convenience of logarithms becomes more apparent in the next example.

EXAMPLE:
Multiply 8.624 by 0.02853 by 26.42

log. 8.624	=	0.9357
log. 0.02853	=	$\bar{2}$.4553
log. 26.42	=	1.4219
		0..8129

The three mantissas add up to 1.8129, that is, +1 is carried over to add to the characteristics, so giving a total of 0 (remember that the logarithm of 0.02853 is really −2 + 0.4553). The anti-logarithm tables give 6500 for 0.8129 and from Table A3.1(ii) the correct answer is <u>6.500</u>.

A3.2 Division

To divide, subtract the logarithms to obtain the logarithm of the quotient, for example, 39 ÷ 13

log. 39	=	1.5911
log. 13	=	1.1139
log. 39/13	=	0.4772

antilog 0.4772 = 3.0 which we know to be correct.

EXAMPLE:
Divide 0.862 by 0.03142

log. 0.8620	=	$\bar{1}.9355$
log. 0.03142	=	$\bar{2}.4972$
log. of quotient	=	1.4383

(note with regard to the characteristics – subtracting –2 from –1 gives –1 – (–2), i.e. –1 + 2, i.e. +1)(A4.1.4)

antilog 1.4383 = 27.44

EXAMPLE:
In Section 3.4.2 a calculation is made of

$$\frac{19.635 \times 0.0008781}{10^6}$$

giving an answer of 0.01724×10^{-6}. Check this using logarithms.

log. 19.635	=	1.2931	(note: log. 19.630 = 1.2930, log. 19.640 = 1.2932, therefore the log. of 19.635 can be estimated as 1.2931)
log. 0.0008781	=	$\bar{4}.9435$	
log. of numerator	=	$\bar{2}.2366$	
log. 10^6	=	6.0000	
log. of quotient	=	$\bar{8}.2366$	

∴ quotient = antilog $\bar{8}.2366$ = 1.724×10^{8}

(note how easily powers of 10 fit into the logarithm system).

In this case the logarithm $\bar{8}.2366$ can be written as two separate quantities +0.2366 and –8.0000 and when the anti-logarithm is taken this becomes 1.724×10^{-8} or in the terms given in the example, 0.01724×10^{-6}.

A3.3 Powers and Roots

To calculate the value of the power of a number, multiply its logarithm by the exponent to obtain the logarithm of the answer.

To calculate the value of the root of a number, divide its logarithm by the exponent to obtain the logarithm of the answer. For example, find by logarithms the value of 2^3

$$\log. 2 = 0.3010$$
$$\log. 2^3 = 0.3010 \times 3 = 0.9030$$

antilog 0.9030 = 7.998 which is very nearly the correct answer of 8.0.

EXAMPLE:
Calculate 0.08431^5

$$\log. 0.08431 = \bar{2}.9259$$
$$\log. 0.08431^5 = \bar{2}.9259 \times 5$$

Never forget that the characteristic is negative while the mantissa is positive. Deal with the latter first:

$$0.9259 \times 5 = 4.6295$$

giving a positive figure of 4 to carry over into the characteristic which is also multiplied by 5:

$$-2 \times 5 = -10$$

add in the 4 gives

$$-10 + 4 = -6$$

combined with the mantissa gives $\bar{6}.6295$, the antilog of which is

$$\underline{0.000004261 \text{ or } 4.261 \times 10^{-6}}$$

EXAMPLE:
Calculate $\sqrt[3]{0.4321}$

$$\log. 0.4321 = \bar{1}.6356$$

$$\log. \sqrt[3]{0.4321} = \frac{\bar{1}.6356}{3}$$

Again, the characteristic and mantissa must be treated separately. The characteristic is always a whole number and to keep it so we manipulate the calculation as follows:

$\bar{1}.6356$ is the same as $-1 + 0.6356$ or $-3 + 2.6356$,

i.e. the characteristic has been increased by the smallest number so that it becomes exactly divisible by 3. In this case 2 has been subtracted from the -1 to make -3. By adding 2 to the mantissa no change has been made in the value of the complete logarithm and it is now divisible by 3:

$$\frac{-3 + 2.6356}{3} = -1 + 0.8785$$

written as $\bar{1}.8785$, the antilog of which is 0.7560, i.e.

$$\underline{\sqrt[3]{0.4321} = 0.7560}$$

There is another way of handling a negative characteristic which is not divisible exactly by the exponent, this is to convert the whole logarithm to a negative number, carry out the division required, and finally convert back to normal logarithmic form so that anti-logarithm tables can be used. The above example is re-calculated in this way:

$$\log. 0.4321 = \bar{1}.6356 = -1 + 0.6356 = -0.3644$$

this is no longer in characteristic plus mantissa form but is a single negative number which is divisible by any other number, so

$$\frac{-0.3644}{3} = -0.1215$$

Re-converted, this becomes $-1 + 0.8785$ or $\bar{1}.8785$, the antilog of which is 0.7560.

Appendix 4

MATHEMATICS

A4.1 Some Reminders about Arithmetic

A4.1.1 Fractions

"Numerator" above the line, "denominator" below. A fraction is unchanged in value if both numerator and denominator are multiplied or divided by the same number, e.g.

$$\frac{8}{12} \text{ has the same value as } \frac{8 \div 4}{12 \div 4}\text{, i.e. } \frac{2}{3}$$

usually written as cancellation:

$$\frac{\overset{2}{\not 8}}{\underset{3}{\not 12}} = \frac{2}{3}$$

$$\text{also } \frac{8}{12} \text{ has the same value as } \frac{8 \times 5}{12 \times 5}\text{, i.e. } \frac{40}{60}$$

It is not permissible to add or subtract the same number to or from both numerator and denominator, e.g.

$$\frac{8}{12} \text{ does not have the same value as } \frac{8+5}{12+5} \text{ or } \frac{8-5}{12-5}$$

Division by a fraction: invert and multiply, e.g.

$$\frac{5}{12} \div \frac{2}{3} \text{ becomes } \frac{5}{12} \times \frac{3}{2} = \frac{5 \times 3}{12 \times 2} = \frac{15}{24}$$

With a fraction in the denominator, e.g.

$$\frac{8}{\frac{5}{12}} \text{ becomes } 8 \times \frac{12}{5} = \frac{96}{5}$$

A4.1.2 Exponents

When a number is multiplied by itself it is said to be ***squared*** or raised to the power of 2, i.e. $x \times x = x^2$. The figure 2 is called the ***exponent*** or ***index***.

Similarly

$x \times x \times x = x^3$ (*x* cubed)

$x \times x \times x \times x = x^4$ (*x* to the fourth power)

etc.

The square root, cube root, fourth root, etc. of a number is that quantity which when multiplied by itself the appropriate number of times, produces the number. The square root of x is written $\sqrt{x}$ and $\sqrt{x} \times \sqrt{x} = x$. Similarly for the cube root of x, $\sqrt[3]{x} \times \sqrt[3]{x} \times \sqrt[3]{x} = x$ and $\sqrt[4]{x} \times \sqrt[4]{x} \times \sqrt[4]{x} \times \sqrt[4]{x} = x$, etc.

These roots however may be expressed with a fractional exponent instead of the root sign ($\sqrt{\ }$):

$$\sqrt{x} = x^{1/2}, \quad \sqrt[3]{x} = x^{1/3}, \quad \sqrt[14]{x} = x^{1/14}$$

Negative Exponents:

When a quantity is raised to a negative power, this is equivalent to it forming the denominator of a fraction:

x^{-2} is the same as $\frac{1}{x^2}$ or $1 \div x^2$

and similarly

$$x^2 \text{ is the same as } \frac{1}{x^{-2}} \text{ or } 1 \div x^{-2}$$

and we can change between division and multiplication by changing the sign of the exponent, e.g.

$$\frac{y}{x^3} = yx^{-3} \text{ , i.e. } y \div x^3 \text{ is the same as } y \times x^{-3}$$

$$\frac{y}{x^{-3}} = yx^3 \text{ , i.e. } y \div x^{-3} \text{ is the same as } y \times x^3$$

Rules for Multiplication of Exponents:

(1) When powers of the same quantity are multiplied together, the exponents are added:

$$x^2 \times x^3 = (x \times x) \times (x \times x \times x) = x^{2+3} = x^5$$

$$x^{1/2} \times x^{1/2} = x^{1/2+1/2} = x^1 = x \text{ (also proving that } \sqrt{x} \times \sqrt{x} = x)$$

$$x^{1/2} \times x^{-1/2} = x^{1/2-1/2} = x^0 = 1$$ (also proving that $(\sqrt{x}/\sqrt{x}) = 1$ and note that quantities raised to the power 0 are all equal to 1).

(2) When powers of the same quantity are divided, the exponent of the divisor (denominator) is subtracted from that of the dividend (numerator):

$$x^3 \div x^2 = \frac{x \times x \times x}{x \times x} = x^{3-2} = x^1 = x$$

$$x^2 \div x^3 = \frac{x \times x}{x \times x \times x} = x^{2-3} = x^{-1} = \frac{1}{x}$$

$$x^{1/2} \div x^{1/4} = x^{1/2-1/4} = x^{1/4} \quad (= \sqrt[4]{x})$$

Rule (2) is of course, rule (1) modified for negative exponents (for example, $x^3 \div x^2$ is the same as $x^3 \times x^{-2}$ and by rule (1) $x^3 \times x^{-2} = x^{3+(-2)} = x$.

(3) A product of two or more factors can be raised to a power by raising each factor to that power:

$$(x.y.z)^3 = x^3 \times y^3 \times z^3 ,$$

generally, $(xyz)^m = x^m . y^m . z^m$

(4) A power of a power is obtained by multiplying the two exponents:

$$(x^2)^3 = (x \times x) \times (x \times x) \times (x \times x) = x^{2\times 3} = x^6$$

$$(\sqrt{x})^4 = \sqrt{x} \times \sqrt{x} \times \sqrt{x} \times \sqrt{x} = x^2$$

or using fractional exponents:

$$(\sqrt{x})^4 = (x^{1/2})^4 = x^{1/2 \times 4} = x^2$$

Powers and Roots of Fractions:
The exponent operates on both numerator and denominator:

$$\left(\frac{x}{y}\right)^2 = \frac{x^2}{y^2} \qquad \pi\left(\frac{d}{2}\right)^2 = \frac{\pi d^2}{4}$$

$$\sqrt{\frac{x}{9}} = \frac{\sqrt{x}}{3}$$

or using fractional exponents:

$$\left(\frac{x}{9}\right)^{1/2} = \frac{x^{1/2}}{9^{1/2}} = \frac{x^{1/2}}{3}$$

A4.1.3 Reciprocals

A reciprocal is a number so related to another that their product is unity, thus the reciprocal of any number is obtained by dividing that number into 1, e.g. the reciprocal of 8 is 1/8 so that 8 x 1/8 = 1, the reciprocal of 0.1 is 1/0.1 = 10.

A4.1.4 Plus and Minus

Positive quantities are indicated by +, and negative quantities are indicated by –.

$8 + 5 = 13$	a term with no sign is assumed +
$8 + (-5) = 8 - 5 = 3$	a + sign outside a bracket causes no sign changes when the brackets are removed
$5 - 8 = -3$	when the – quantity is numerically greater than the + quantity the + is subtracted from the – and the answer is –
$5 - (-8) = 5 + 8 = 13$	a – sign outside a bracket changes all signs within the brackets when the latter are removed
$8 \times 5 = 40$	i.e. + x + = +
$-8 \times 5 = -40$) $8 \times -5 = -40$)	– x + = –
$-8 \times -5 = 40$	– x – = +
$\frac{-8}{5} = \frac{8}{-5} = -1.6$	$\left(\frac{8}{-5} \times \frac{-1}{-1} = \frac{-8}{5}\right)$
$\frac{-8}{-5} = 1.6$	$\left(\frac{-8}{-5} \times \frac{-1}{-1} = \frac{8}{5}\right)$

A4.1.5 Percentages

Per cent is derived from the Latin "per centum", "in or to every hundred" and in our use it means just that. Instead of expressing fractions with different denominators, often having no easily recognisable correlation between them (e.g. 3/7 is exactly the same as 39/91, but it is not immediately obvious), we express fractions to a denominator of 100 and label them *per cent*, symbol %. Thus both 3/7 and 39/91 may be expressed as 42.86 per cent because each is equal to 42.86/100, the percentage value being obtained by multiplying by 100. Conversely to bring a percentage back to a fraction, divide by 100. Examples follow:

(i) What is 30% of 1 metre?

$$30\% \text{ of 1 metre (100 cms)} = \frac{30}{100} \times 100 = \underline{30 \text{ cms}}$$

(ii) A voltage varies within the range 6 V ± 25%. What are the maximum and minimum voltages?

$$25\% \text{ of } 6\text{ V} = \frac{25}{100} \times 6 = 1.5\text{ V}$$

∴ max. and min. voltages are (6 + 1.5) and (6 – 1.5), i.e. <u>7.5 and 4.5 volts</u>

(iii) A 12-volt car battery voltage actually varies between 14 V on charge and 10.8 V when discharged. Express the variation as a percentage of the nominal voltage.

$$\text{Variation of voltage} = (14 - 12)\text{V to } (10.8 - 12)\text{V}$$

$$= +2 \text{ to } -1.2\text{ V}$$

∴ Variation as a percentage

$$= +\left(\frac{2}{12} \times 100\right) \text{ to } -\left(\frac{1.2}{12} \times 100\right)$$

$= +16.67\%$ to -10%

(iv) A 330-ohm resistor has a tolerance of ± 20%. What are the extreme values?

$$20\% \text{ of } 330 \text{ ohms} = \frac{20}{100} \times 330 = 66 \text{ ohms}$$

∴ Extreme values are (330 + 66) and (330 – 66), i.e. 396 and 264 ohms

A4.2 Simple Equations

Much of mathematics revolves around the "equals" (=) sign, defined by the dictionary as "the same in number, size, value, degree, etc.". Mathematical equations have two parts separated by the equals sign and for convenience we will refer to the left-hand side (l.h.s.) and the right-hand side (r.h.s.). When an equation has been set up, any operation which does not change the equality between the two sides is permissible, any which does destroy the equality is not [unless the equals sign is changed for example, to ≈ (approximately equal to) or ≠ (not equal to)]. Common operations are:

multiplication of both sides by the same quantity

division of both sides by the same quantity

adding the same quantity to both sides

subtracting the same quantity from both sides

squaring both sides

taking the square root of both sides.

Ohm's Law (Section 3.1) gives an example of the simple equation:

$$I = \frac{V}{R} \quad \dots\dots\dots\dots\dots\dots\dots\dots \quad (1)$$

(I = current through a circuit, V = voltage applied and R = resistance of circuit), to find V in terms of I and R, multiply both sides of equation (1) by R:

$$I \times R = \frac{V}{R} \times R$$

and R cancels on the r.h.s. having

$$I \times R = V$$

(I x R can also be written I . R or IR)

$$\therefore \quad \underline{V = IR} \quad \dots\dots\dots\dots\dots\dots\dots\dots \quad (2)$$

Similarly to find R in terms of I and V, divide both sides of equation (2) by I:

$$\frac{V}{I} = \frac{\cancel{I}R}{\cancel{I}} = R, \text{ i.e. } \underline{R = \frac{V}{I}}$$

Cross-multiplication:
If

$$\frac{a}{b} = \frac{c}{d}$$

then by cross-multiplication, i.e.

$\frac{a}{b}$ multiplied by $\frac{c}{d}$ (a → d, b → c)

$$\underline{ad = bc}$$

equivalent to but quicker than multiplying both sides by bd, thus:

$$\frac{a}{b} = \frac{c}{d} \quad \therefore \frac{abd}{b} = \frac{cbd}{d} \quad \therefore \underline{ad = bc}$$

We can also change over numerators and denominators on both sides for if bc = ad, then dividing both sides by ac gives

$$\underline{\frac{b}{a} = \frac{d}{c}}$$

A4.2.1 Brackets

A term outside of a bracketed quantity operates on each term within the brackets when the latter are removed, e.g. 7(6a + 2b) becomes 42a + 14b when the brackets are removed. When the first bracket contains more than one term, the second bracket terms are multiplied by each of them, e.g.

$$(a + b)(c + d) = a(c + d) + b(c + d) = ac + ad + bc + bd$$

$$(a - b)(c + d) = a(c + d) - b(c + d) = ac + ad - bc - bd$$

The following occur frequently:

$$(a + b)^2 = (a + b)(a + b) = a(a + b) + b(a + b)$$

$$= a^2 + 2ab + b^2$$

$$(a - b)^2 = (a - b)(a - b) = a(a - b) - b(a - b)$$

$$= a^2 - 2ab + b^2$$

$$(a + b)(a - b) = a(a - b) + b(a - b) = a^2 - b^2$$

Reversing the process, quantities having a common factor may be bracketed with the factor placed outside, e.g. $6a^2 + 2ab$ has the common factor 2a for $(6a^2/2a) = 3a$ and $(2ab/2a) = b$, thus $6a^2 + 2ab$ could be written as $2a(3a + b)$.

A4.2.2 The symbol $\propto$

Shorthand for "is proportional to" or "varies as" is the symbol $\propto$. An example which we know to be true is that the weight of a newspaper is proportional to the number of pages. This follows because if one page has a weight w, then two pages weigh 2w and so on. It might also be found that for a certain person the reading time is proportional to the number of pages. Thus for this newspaper weight (w) $\propto$ number of pages (n). Time to read (t) $\propto$ number of pages (n).

These are not mathematical equations because we have no knowledge of actual weights, times or numbers of pages. But when one relationship between these is known, then an equation can be set up. Suppose 10 pages weigh 50 grammes and take one hour to read, these are known relationships enabling us to calculate others, e.g. the weight (w_2) and time to read (t_2) of 80 pages (n_2), for

$$\frac{w_1}{w_2} = \frac{n_1}{n_2} \quad \text{and} \quad \frac{t_1}{t_2} = \frac{n_1}{n_2}$$

where w_1, n_1 and t_1 are in the known relationship. Substituting:

$$\frac{50}{w_2} = \frac{10}{80} \quad \text{and} \quad \frac{60}{t_2} = \frac{10}{80} \quad \text{(t in minutes)}$$

Cross-multiplying in both cases:

$$10w_2 = 50 \times 80, \therefore w_2 = \underline{400\text{ g}}$$

$$10t_2 = 60 \times 80, \therefore t_2 = 480\text{ mins} = \underline{8\text{ hours}}$$

When one quantity (x) is inversely proportional to another (y), we write

$$x \propto \frac{1}{y} \qquad \text{or } y \propto \frac{1}{x}$$

and the equation becomes

$$\frac{x_1}{x_2} = \frac{y_2}{y_1}$$

and again, given any three, the fourth can be calculated.

Also when one quantity (x) is proportional to another (y) and also to a third (z):

$$\frac{x_1}{x_2} = \frac{y_1 z_1}{y_2 z_2}$$

or if inversely proportional to z,

$$\frac{x_1}{x_2} = \frac{y_1 z_2}{y_2 z_1}$$

Using the constant, k:
Once the actual relationship between the quantities, say, w and n, is known, it is then possible to change from the $\propto$ sign to = to give an equation linking them by inserting a factor (k) as for example:

If 10 pages weigh 50 g, 1 page weighs 50/10 = 5 g and this gives a value for k of 5 (grammes per page).

Since w $\propto$ n

$\therefore \quad$ w = kn

and if, as in this example, k = 5

$$w = 5n$$

where w is in grammes, from which w_2 in the above example would simply be calculated:

$$w_2 = 5n_2 = 5 \times 80 = \underline{400.}$$

Appendix 5

GEOMETRY

Except for arithmetic, geometry is probably the most used branch of mathematics in daily life. In early days it was the science of earth measurement (Greek, "geo", earth; "metron", measure), but now it embraces all *plane* measurements (rectangles, circles, angles, etc. – also termed "two-dimensional", e.g. length x breadth) and *solid* geometry (cubes, spheres, etc. – also termed "three-dimensional, e.g. length x breadth x height). In first studies of electronics we need mainly to be able to calculate the volume of a cylinder for this is in fact the shape of a piece of wire, and to do this we must become conversant with the circle first.

A5.1 The Circle

If we take a line such as OR in Fig. A5.1(i) and rotate it about an imaginary pivot at point O through a quarter of a turn (ii), then the line has moved through 90 degrees (90°) which is also called a *right-angle*, colloquially OR and OR_1 form a "square corner". Note that OR has been turned in an anti-clockwise direction, this is the convention used almost exclusively in electronics. Turning OR through a further 90° (iii) brings it into the position OR_2, i.e. 180° shift and the end of the line remote from O has traced out a *semicircle*. Continuing the process through a further 180° completes the circle which is thus divided into 360°. Degrees may also be subdivided into minutes, with 60 minutes (60′) to 1 degree (1°). As with time (days, hours, etc.) there are no plans to change with metrication.

Fig. A5.1(iv) shows a complete circle as would be drawn by rotation of OR about the point O. OR is known as the *radius* (r) and any line drawn across the circle through the centre point O is a *diameter* (d). Clearly d = 2r and for any circle, the length of the circumference (imagine the circle

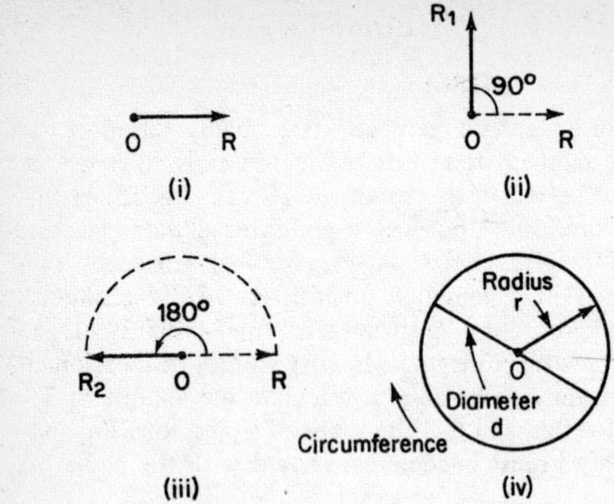

Fig. A5.1 The circle.

being cut and pulled out to a straight line) has a fixed relationship to the diameter, known by the Greek letter π (pi), hence length of circumference = $\pi \times d$. The actual value of π is rather untidy because it does not work out as an exact number so we have to choose the degree of accuracy we require for any particular purpose. π to more decimal places than are normally needed has a value of 3.141592654, for most work 3.142 is adequate. A few useful figures based on π are:

$$\pi = 3.14159$$

$$\frac{1}{\pi} = 0.31831$$

$$\sqrt{\pi} = 1.77245$$

$$\pi^2 = 9.86960$$

$$\log \pi = 0.4971^{(A3)}$$

As fractions,

$$\frac{22}{7} \; (= 3.14286)$$

$$\frac{355}{113} \; (= 3.14159292)$$

The last fraction is a sufficiently accurate representation of π for most purposes.

A5.2 Areas

The size of an area is 1 square metre (1 m^2) if it occupies a plane space equivalent to a square of 1 m side.

The area of a *rectangle* (a 4-sided figure, all sides meeting at rt ∠s) is given by length x breadth. If length and breadth are equal, the figure is a square.

The area of a circle is given by πr^2 where r is the radius, and since the diameter, d, = 2r, ∴ r = d/2, the area is also equal to $\pi d^2/4$.

EXAMPLE:
A coin, 2 millimetres thick has a diameter of 3 centimetres. What is the total surface area? [1 centimetre (cm) = 10 millimetres (mm), see Appendix 6]

The total surface area is made up of two circular faces of 3 cm diameter and one rectangular strip (the edge), 2 mm wide and in length equal to the circumference as shown in Fig. A5.2.

$$d = 3 \text{ cm} \qquad \therefore r = 1.5 \text{ cm}$$

$$\text{Area of one face} \quad = \pi r^2 = 3.142 \times 1.5^2$$

Area of both faces $= 2 \times 3.142 \times 1.5^2 = 14.139\ cm^2$

Circumference $= \pi d = 3.142 \times 3$

$\therefore$ Area of edge $= 3.142 \times 3 \times 0.2\ cm^2 = 1.885\ cm^2$

$\therefore$ Total area $= 14.139 + 1.885 = \underline{16.024\ cm^2}$

For convenience, by realising that π occurs in both calculations, we could have brought it out as a common factor thus:

Area of both faces $= \pi \times 2 \times 1.5^2$

Area of edge $= \pi \times 3 \times 0.2$

$\therefore$ Total area $= \pi[(2 \times 1.5^2) + (3 \times 0.2)] =$

$= \pi(4.5 + 0.6)$

$= 3.142 \times 5.1 = \underline{16.024\ cm^2}$

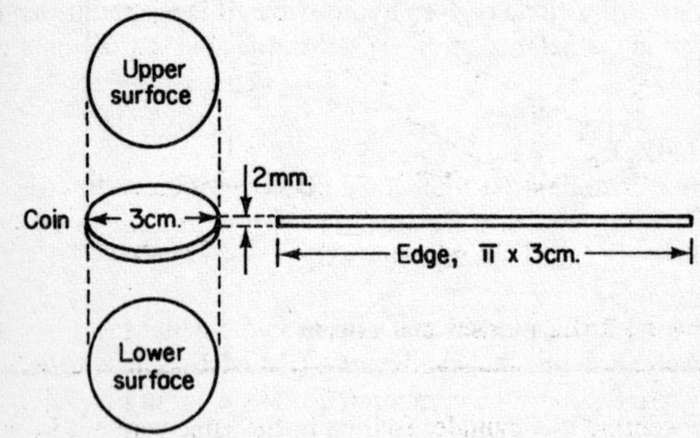

Fig. A5.2 Surface area of a coin.

A5.2.1 Area of cross-section

This is a term frequently met in engineering, a cross-section is formed by imagining a cut being made across the item, usually at right-angles to its length, and viewing the cut end. We are mainly concerned with cross-sections which are rectangular or circular for which the areas are calculated as above.

A5.3 Volumes

Volumes bring us to the 3-dimensional and the amount of a volume is 1 cubic metre (1 m^3) if it occupies a bulk equivalent to that of a cube of 1 m side.

A *right prism* is a solid figure with similar rectangular ends and parallel sides as shown in Fig. A5.3. Its volume is calculated from length x breadth x height or equally from area of base x height and obviously it does not matter which surface is taken as the base.

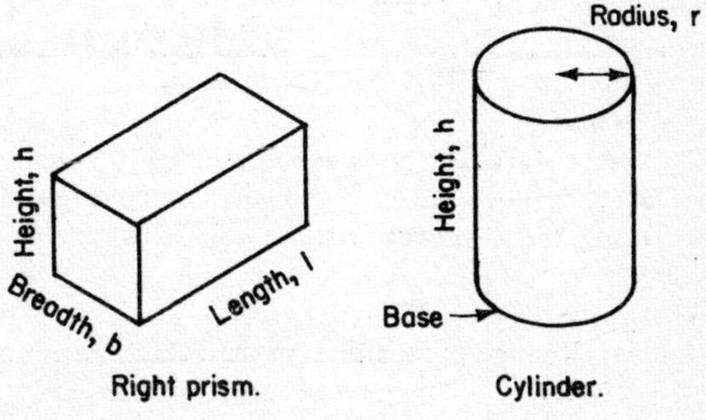

Fig. A5.3 Right prism and cylinder.

The volume of a cylinder follows in the same way but in this case the base is one of the end circular surfaces thus:

$$\text{Volume of cylinder} = \text{area of base x height} = \pi r^2 h$$

For a wire therefore,

Volume = area of cross-section x length

EXAMPLE:
What is the area of cross-section of a 10 m length of 4 mm (diameter) copper wire and what is its weight if copper weights 8.89 grammes per cubic centimetre (g/cm^3)?

$d = 4$ mm $\qquad r = 2$ mm

Area of cross-section $= \pi r^2 = \pi \times 2^2 = 3.142 \times 4$

$= 12.57\ mm^2$ or $0.1257\ cm^2$

(note that if there are 10 mm to 1 cm, there must be $10 \times 10\ mm^2$ to $1\ cm^2$ – see below).

Volume = area of cross-section x length

$= 0.1257 \times 1000 = 125.7\ cm^3$

$\therefore$ Weight $= 125.7 \times 8.89$ g $= 1117$ g $= 1.117$ kg

Unit conversion:
When converting area or square measure units, the linear conversion factor is squared(A4.1.2) and when convertng volume measure units, the linear conversion factor is cubed. For example:

Linear conversion factor: 1000 mm = 1 m

Area or square measure conversion factor:
1000^2 sq. mm (mm^2)
= 1 sq. m (m^2)

Volume conversion factor: 1000^3 cu. mm (mm^3)
= 1 cu. m (m^3)

and as shown in the example above 10 mm = 1 cm, $\therefore$ 100 mm^2 = 1 cm^2 and 1000 mm^3 = 1 cm^3.

Appendix 6

MULTIPLES AND SUB-MULTIPLES OF UNITS

One of the great advantages of the metric system is that, whatever the "unit", for example a length of 1 metre or weight of 1 gramme, larger or smaller quantities are all described by the same range of prefixes. This also applies throughout electronics and as we learn of new units this standard range of prefixes will be used. Table A6.1 and the examples which follow will make this clear.

Some of the multiples and/or sub-multiples will be found to occur more frequently than others depending mainly on the convenience of the size of the unit and how usage has grown traditionally. As an example, one gramme is such a small weight for shopping that a kilogramme (1000 grammes) is more useful. Half a kilogramme however is more likely to be stated as 500 grammes rather than 5 hectogrammes and in fact for this purpose hectogrammes and decagrammes are much less frequently used than grammes and kilogrammes.

The same holds good in electronics and those prefixes which are likely to occur most frequently in basic electronics and are recommended for common use appear in Table A6.2. 10^2 hecto, 10^1 deca and 10^{-1} deci are rarely used.

The main units used in everyday life are:

length – metre (m)	(about 39 inches)
weight – gramme (g)	(approx. 454 g = 1 lb)
capacity – litre (ℓ)	(about 1¾ pints)

EXAMPLE:
How many centimetres (cm) are there in 1 kilometre (km)?

Relative to 1 metre $\frac{\text{km}}{\text{cm}} = \frac{10^3}{10^{-2}} = 10^{3+2} = \underline{10^5\ (100{,}000)}$

(dividing by 10^{-2} is the same as multiplying by 10^2 – see Appendix 2).

EXAMPLE:
A power-station generates 1 megawatt of power. How many kilowatts is this equivalent to (the unit is 1 watt)?

Relative to 1 watt $\frac{\text{MW}}{\text{kw}} = \frac{10^6}{10^3} = 10^{6-3} = \underline{10^3\ (1{,}000)}$

Table A6.1 Multiples and Sub-Multiples of Units

Multiplication Factor (shortened form – see Appendix 2)	10^9	10^6	10^3	10^2	10^1	10^{-1}	10^{-2}	10^{-3}	10^{-6}	10^{-9}	10^{-12}
Prefix	giga	mega	kilo	hecto	deca	deci	centi	milli	micro	nano	pico
Symbol	G	M	k	h	da	d	c	m	μ*	n	p

* This letter is from the Greek alphabet, called MU, pronounced "mew".

Table A6.2 Units in Electronics

Quantity	*Unit*	*Symbol*	*Multiples and sub-multiples recommended for common use*										
			G	*M*	*k*	*h*	*da*	*d*	*c*	*m*	*μ*	*n*	*p*
Length	metre	m			km				cm	mm	μm	nm	
Mass	gramme	g			kg					mg	μg		
Time	second	s			ks					ms	μs	ns	
Electric current	ampere	A			kA					mA	μA	nA	pA
Frequency	hertz	Hz	GHz	MHz	kHz								
Energy, work	joule	J		MJ	kJ					mJ			
Power	watt	W			kW					mW	μW	nW	pW
Magnetic flux	weber	Wb								mWb			
Magnetic flux density	tesla	T								mT			
Electric charge	coulomb	C			kC						μC	nC	pC
Electric potential, PD, EMF	volt	V			kV					mV	μV		

(table continued)

Quantity	*Unit*	*Symbol*	*Multiples and sub-multiples recommended for common use*										
			G	*M*	*k*	*h*	*da*	*d*	*c*	*m*	*μ*	*n*	*p*
Capacitance	farad	F								mF	μF	nF	pF
Self and mutual inductance	henry	H								mH	μH	nH	pH
Conductance	siemen	S			kS					mS	μS		
Resistance	ohm	Ω		MΩ	kΩ						μΩ		
Impedance	ohm	Ω		MΩ	kΩ					mΩ			

[Note that symbols for units derived from personal names are in capital letters.]

Appendix 7

CIRCUIT SYMBOLS

Some of the more commonly used electronic circuit symbols are given below. These are almost universal but countries have their own preferences in some cases. Some variations are also given.

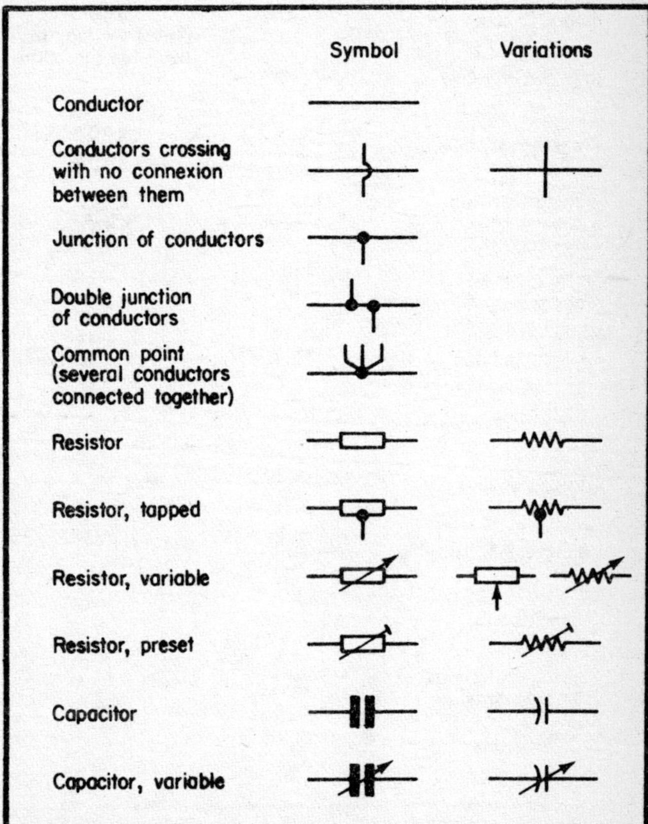

Continued.......

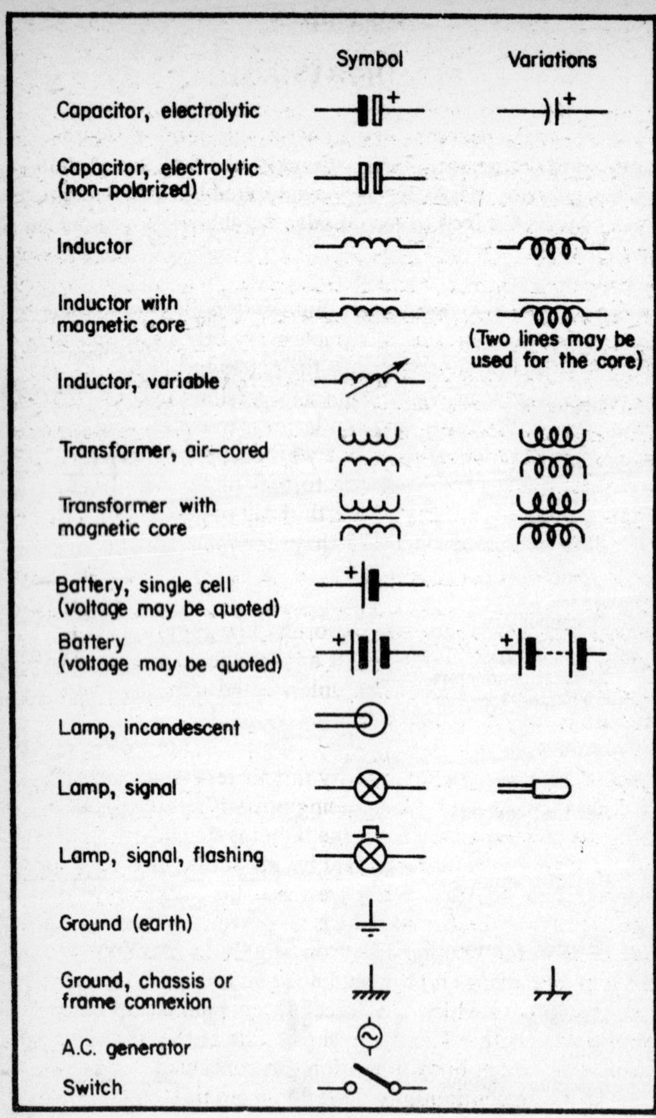

Use of a few of the above in a circuit diagram is illustrated by Fig. 3.2.

Appendix 8

ORBITS

We increasingly become familiar with this term as space exploration progresses. The Earth orbits the Sun, the Moon orbits the Earth, space capsules have orbited both Moon and Earth. But first a look at two fundamental laws.

Gravity:
Sir Isaac Newton in the sixteen hundreds published his law which said that every object attracts every other and that the force of attraction increases with their sizes but decreases with the distance between them. Thus a man falling out of an aeroplane attracts the Earth and the latter attracts him, but we know from experience which one wins because of the enormous mass of the Earth relative to that of the man, one simply could not imagine the Earth rising to meet the man. We call the attraction of the Earth on the man, *gravity*.

Motion:
Newton also gave us another important law about motion by stating that all bodies stay at rest or if moving, continue to do so uniformly in a straight line unless acted upon by some external force.

Thus, if there were neither gravity nor air resistance, a bullet fired into the air would keep going but with each of these having its own separate effect, the bullet is slowed by the air and eventually falls to the ground by the action of gravity as shown in Fig. A8.1(a). When we reach the nothingness of space, far above the Earth, there is no air to create resistance and therefore no braking effect on a body in motion, so it just goes on moving and it would do so in a straight line were it not for gravity which is a force always pulling the body towards the Earth. The figure shows this at (b), that is, a bullet or any other body in motion can continue to circle the Earth by being continuously deflected from its straight line of flight by the pull of gravity. If the force of gravity were to be suddenly removed, the bullet would fly off in a straight line as does a stone released from a sling.

The gravitational pull required to maintain an orbit depends on the speed of the bullet; too great a pull (closer to Earth) for a given speed will cause the bullet or object to fall downwards; too little will produce an increasing orbit until eventually the bullet is lost. Thus distance from Earth and speed of a space capsule must be accurately calculated for the capsule to fall into a particular orbit. Once orbiting, no driving power is required, this is only needed to change things, for example, move out towards another planet or in towards the Earth or to change to a smaller or larger orbit.

The electron orbits its atom in exactly the same way, controlled by the same principles. Of particular interest is the fact that an electron can be shifted out of its orbit by an external force.

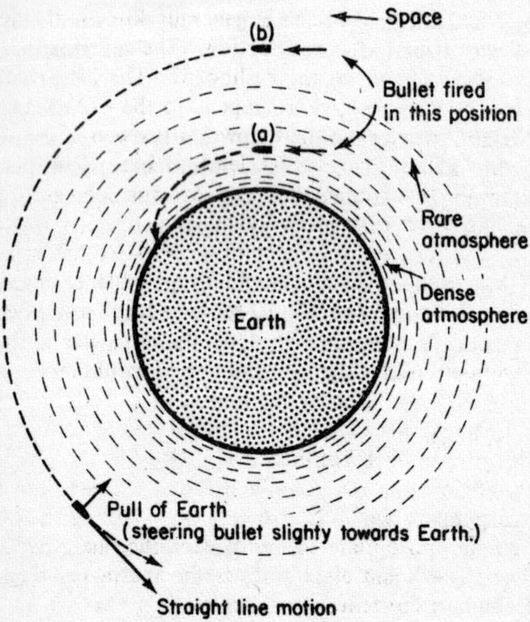

Fig. A8.1 Bullet in orbit around the Earth.

Appendix 9

WORK AND POWER

The dictionary will show that the terms in the title have many different meanings and applications in everyday language, so to avoid ambiguity the physicist has to define them more carefully. Included also are other quantities which need explanation, for example, *work* contains a component *force* which itself contains *acceleration*. The latter two are therefore discussed first.

A9.1 Acceleration

When a car covers a distance at a rate of, say, 80 km per hour (about 50 m.p.h.) this is generally known as its speed or *velocity* if in a known direction. The speed is indicated by the speedometer on the dashboard. The driver understands acceleration as what happens to the car when the accelerator pedal is pushed down, its speed increases. Whenever speed changes, acceleration has taken place (positive for increasing speed, negative for decreasing speed, also called "deceleration"). Technically the manufacturer might have quoted the acceleration as "0–80 km/h in 10 seconds" meaning that it should be possible to reach a velocity of 80 km/h within 10 seconds from rest. Because acceleration is rate of change of velocity, the average acceleration during the 10 second period would be

$$\frac{80\ \text{km/h}}{10\ \text{secs}} = 8\ \text{km/h/s}.$$

This unit is appropriate to motor cars, the one especially used in physics and electronics is the metre per second per second, in shortened form 1 m/s^2.

EXAMPLE:
A car has an acceleration of 1 m/s^2. How long will it take to accelerate from 20 to 50 km/h if the acceleration is constant?

First bring the car speeds to metres per second (m/s)

$$20 \text{ km/h} = \frac{20}{60 \times 60} \text{ km/s} = \frac{20 \times 1000}{60 \times 60} \text{ m/s}$$

$$= 5.556 \text{ m/s}$$

$$50 \text{ km/h} = \frac{50 \times 1000}{60 \times 60} \text{ m/s} = 13.889 \text{ m/s}$$

$\therefore$ Change in speed required

$$= 13.889 - 5.556 = 8.333 \text{ m/s}$$

and if speed changes 1 m/s each second then time to change speed

$$= \frac{8.333 \text{ m/s}}{1 \text{ m/s}^2} = \underline{8.333 \text{ seconds}}$$

A9.2 Force

We cannot say exactly what force is but we can observe what it does. Force can be exercised by the push or pull of muscular action or by the efforts of machinery to move an object, from which arises the statement that a force "is that which causes a change in the velocity of a body". The unit of force, being defined by what it does, is that force which when applied to a mass of 1 kilogramme, gives it an acceleration of 1 m/s^2. This unit is named the newton (symbol N – after

Sir Isaac Newton, the English physicist and mathematician). Hence a force of, say, 3 N applied to a mass of 1 kg will produce an acceleration of 3 m/s^2 or equally when applied to a mass of 3 kg will produce an acceleration of 1 m/s^2 (no other forces, e.g. friction, must be acting). The general formula is

$$\text{Force (newtons)} = \text{mass (kg)} \times \text{acceleration } (m/s^2)$$

EXAMPLE:
The force of gravity on a mass of 1 kg is approximately 9.8 N. If such a mass is allowed to fall freely from a height, unhampered by air resistance, what velocity would it reach after 5 seconds?

$$F\ (N) = m\ (kg) \times a\ (m/s^2)$$

$$\therefore 9.8 = 1 \times a$$

$$\therefore a = 9.8\ m/s^2$$

$$\therefore \text{Velocity after 5 seconds} = 9.8 \times 5 = \underline{49.0\ m/s}$$

Note: this of course is a theoretical consideration, in practice air has a braking effect on a body falling through the atmosphere and reduces its acceleration. We deliberately cause this to happen with a parachute where an almost constant velocity rather less than 8 m/s is achieved.

A9.3 Work

When a force moves anything, work is done, it is defined as "force x distance through which the force acts". The international unit of work is the joule (symbol J, after James Prescott Joule, an English physicist) such that one joule is the work done when a force of one newton moves a distance of one metre, i.e.

$$\text{Work done, W (joules)} = \text{Force (newtons)} \times \text{distance (metres}$$

EXAMPLE:
A man exerts a pull of 2 N on a trolley and moves it 8 m.
How much work has he done?

$$\text{Work done} = \text{force} \times \text{distance}$$

$$= 2\text{ N} \times 8\text{ m} = \underline{16\text{ joules}}$$

EXAMPLE:
A bucket of sand of weight 10 kg is lifted 8 m. How much work is done? (assume the pull of gravity on 1kg to be 9.81 N).

The force necessary to lift the weight must overcome the effect of gravity which is a force of 9.81 N per kg, i.e. a total of $9.81 \times 10 = 98.1$ N.

Then, work done in lifting the sand

$$= 98.1 \times 8 = \underline{784.8\text{ joules}}$$

(This force in fact only balances that of gravity, in practice a slightly greater force must be applied.)

A9.4 Energy

Energy is the capacity for doing work and it is measured in the same unit, i.e. the joule. Many different forms of energy exist and they are all capable of doing work by exercising a force through a distance, e.g.

(i) a moving body (such as the football in Section 1.2.3) – this is known as *kinetic* energy (from the Greek "due to motion");
(ii) a body raised to a height;
and
(iii) a compressed or stretched spring – both known as *potential* energy (from the Latin "to be able");

(iv) chemical and atomic, as fuels which usually produce energy in the form of heat. Fuels may be classified by their *calorific values* which is the energy output obtained when the fuel is used, for example, liquid butane gas has a calorific value of about 49 megajoules (49×10^6 joules) per kilogramme, anthracite about 33 MJ/kg (also see the example in the next section for electricity).

A9.5 Power

Power is the rate of doing work or equally the rate of consumption of energy. A reminder of power is the now old-fashioned "horse-power" brought into use when steam engines were invented and were replacing horses. The unit of work was the raising of 550 pounds one foot – if this were accomplished in one second then the power expended was 1 horse-power. A 1 HP engine of course could keep going at this rate, but a horse could not. The modern unit of power is the watt which is a rate of working of 1 joule per second. It may help to get the different units into perspective to know that the equivalent of 1 HP is 746 watts, that is 1 HP = 746 joules of work done per second.

It is important to appreciate the difference between work and energy on the one hand which take no account of time, and power on the other hand which does. Thus two engines may be used to pump out a pool, they will both do the same amount of work but the engine with the greater power will complete the job in less time.

When considering power dissipation in the form of heat, as, for example, in an ordinary electric fire, the joule per second (watt) is rather small and accordingly the kilowatt (kW) is used:

$$1\ \text{kW} = 1000\ \text{W} = 1000\ \text{J/s}$$

typical electric fires having dissipations of 1 – 3 kW.

EXAMPLE:
What is the calorific value of electricity? The unit is the kilowatt-hour (kWh)

$$1\ \text{kWh} = 1000\ \text{Wh} = 1000 \times 60 \times 60 \text{ watt-seconds (i.e. joules)}$$

$$= 1000 \times 60 \times 60 \times 10^{-6}\ \text{MJ}$$

$$= \underline{3.6\ \text{MJ per unit}}$$

Appendix 10

PREFERRED VALUES AND MARKING OF COMPONENTS

A10.1 Tolerances

Tolerance is a measure (usually in percentage terms(A4.1.5)) of the degree to which the actual value may vary from the design or nominal value. When many types of component are manufactured (especially resistors and capacitors) the product will have a range of values varying above and below the intended one. A typical example might be the manufacture of a batch of 15, 1000 ohm composition resistors, producing a range of values set out in numerical order as shown in Fig. A10.1. The figures look bad but it must be appreciated that we are dealing with a type which is very cheap to produce but which cannot be made with great accuracy. However from a batch the manufacturer can select to produce the tolerances required.

Tolerances with no plus or minus signs mean that the actual value may vary above or below the nominal value by as much as the tolerance, e.g. purchasing a 100 ohm, 10% resistor means that the value of the item received can be any between 1000 – (10% of 1000) ohms and 1000 + (10% of 1000) ohms, i.e. between 900 and 1,100 ohms. Similarly a 1000 ohm, 5% resistor would lie between 950 and 1050 ohms. When the tolerances are not the same above and below the nominal value, they are expressed separately, e.g. +20%, –10%.

Referring to Fig. A10.1 the 15 resistors would be measured for resistance value automatically and sorted as shown, this batch producing 2 resistors within 5% tolerance, 5 within 10%, 7 within 20% and 1 with greater than 20%. From the figure it is also clear that any resistor within a certain tolerance must also be within a greater one.

It is obviously easier to produce wide tolerance (low accuracy) resistors, accordingly these are less expensive and good

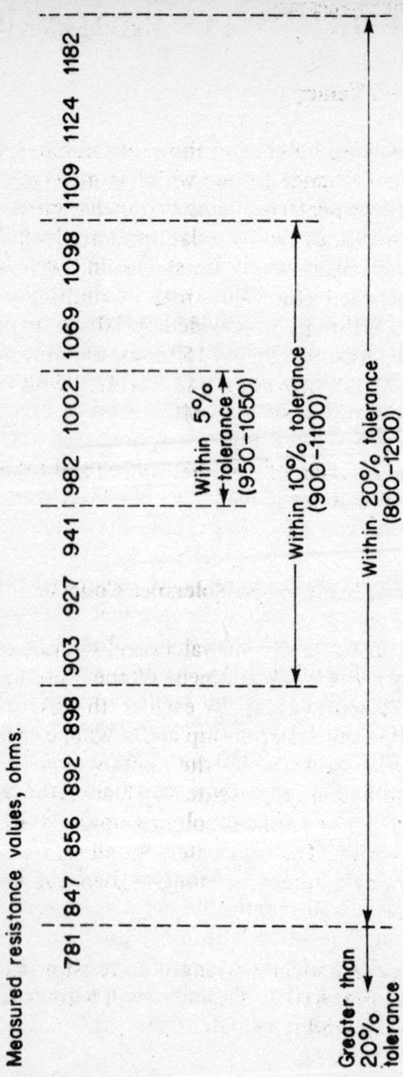

Fig. A10.1 Tolerances in a batch of resistors.

design practice will never specify a close tolerance component when it is not necessary.

A10.2 Preferred Values

The requirement is to reduce the enormous number of values within the practical range to one which is manageable and for this purpose the *preferred value* system has evolved. With due regard to tolerances and overlap, preferred values for resistors and capacitors are those shown in Table A10.1. This is the basic table, the values may be multiplied by a multiple of 10, or if required, divided by 10. As an example, one of the basic resistor values is 15 ohms, the other values obtainable in the 15 range are 1.5 Ω, 150 Ω, 1.5 kΩ, 15 kΩ, 150 kΩ, 1.5 MΩ and possibly 15 MΩ.

An example on selection of preferred values and tolerances is included in Section 3.4.4.

A10.3 Marking (3-colour and Tolerance Codes)

It is sometimes difficult to print values and tolerances on small components, hence the development of the 3-colour code which uses a different colour for each of the digits 0 – 9. Table A10.2 shows the relationship and it will be seen that the digits 2 – 7 are represented by the rainbow colours minus indigo. 0, 1 and 8, 9 are represented by black, brown, grey and white, that is non-rainbow colours moving through the range black to white. The code caters for all values of two figures followed by a number of noughts, hence it is suited to preferred values.

The three colours are usually arranged on resistors and capacitors as shown in Fig. A10.2. Examples will help understanding of how the system works:

(i) Resistance value 3,600 Ω – first figure 3 coded by orange, second figure 6 coded by blue, the subsequent number of noughts is 2, coded by red. Three bands or dots in the order orange, blue, red are therefore the code.

(ii) 82 kΩ – coded by grey, red, orange.

(iii) 13 Ω – coded by brown, orange, black (black indicates no noughts).

(iv) 1.5 MΩ – coded by brown, green, green.

Table A10.1 Preferred Values for Resistors and Capacitors

Percentage Tolerance	*Preferred Values*											
5	10	11	12	13	15	16	18	20	22	24	27	30
10	10		12		15		18		22		27	
20	10				15				22			

Percentage Tolerance	*Preferred Values*											
5	33	36	39	43	47	51	56	62	68	75	82	91
10	33		39		47		56		68		82	
20	33				47				68			

Resistance values are always coded in ohms as shown, capacitance values are coded in picofarads:

(v) 0.033 μF (33,000 pf) – coded by orange, orange, orange.
(vi) 1.0 μF (1,000,000 pF) – coded by brown, black, green.

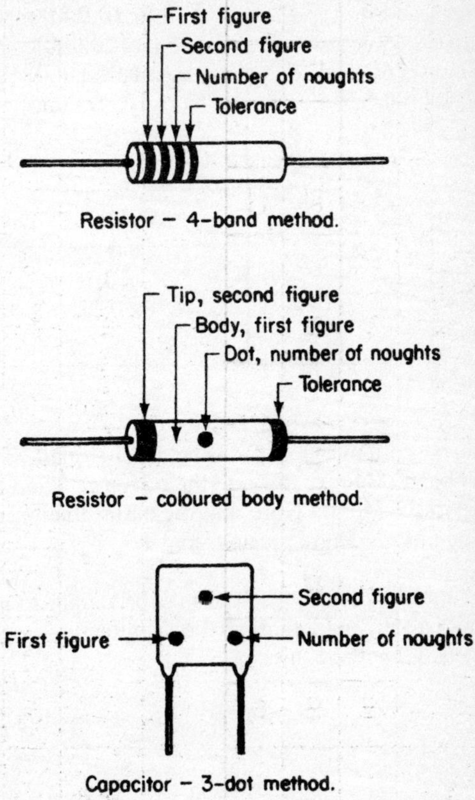

Fig. A10.2 Typical colour coding of resistors and capacitors.

Table A10.2 Colour and Tolerance Code

Colour	*First figure*	*Second figure*	*Number of noughts (multiplier)*	*Tolerance*
Black	0	0	None	
Brown	1	1	1 (x 10)	1%
Red	2	2	2 (x 100)	2%
Orange	3	3	3 (x 1000)	
Yellow	4	4	4 (x 10,000)	
Green	5	5	5 (x 100,000)	
Blue	6	6	6 (x 1,000,000)	
Violet	7	7	7 (x 10,000,000)	
Grey	8	8	–	
White	9	9	–	
			Multiplier	
Gold	–	–	x 0.1	5%
Silver	–	–	x 0.01	10%
None	–	–		20%

Silver and gold are used for an additional mark, forming a fourth band or dot to indicate the tolerance when this is 10% (silver) or 5% (gold). Note that the better tolerance (5%) is indicated by the more precious metal.

This is the standard code, there may be variations in the way components are marked, if so the manufacturer's specification will usually mention this.

Inductance and capacitance are both of major importance in communications for without these we would be denied the wonders of radio altogether. Working in partnership they enable us to select one particular radio transmission out of the hundreds being picked up by an aerial, to sort out the Tower of Babel in effect.

Many other communication developments have at their heart the selective qualities of inductance and capacitance together and this fascinating subject with also a full exploration of the general theory is developed under the title "Alternating Current Theory" in Book 2.

Notes

Notes

Please note overleaf is a list of other titles that are available in our range of Radio and Electronics Books.

These should be available from most good Booksellers, Radio Component Dealers and Mail Order Companies.

However, should you experience difficulty in obtaining any title in your area, then please write directly to the publishers enclosing payment to cover the cost of the book plus adequate postage.

If you would like a copy of our latest catalogue of Radio and Electronics Books then please send a Stamped Addressed Envelope to:–

BERNARD BABANI (publishing) LTD
The Grampians
Shepherds Bush Road
London W6 7NF
England

160	Coil Design and Construction Manual	£2.50
202	Handbook of Integrated Circuits (ICs) Equivalents and Substitutes	£2.95
205	Hi-Fi Loudspeaker Enclosures	£2.95
208	Practical Stereo and Quadrophony Handbook	£0.75
214	Audio Enthusiast's Handbook	£0.85
219	Solid State Novelty Projects	£0.85
220	Build Your Own Solid State Hi-Fi and Audio Accessories	£0.85
221	28 Tested Transistor Projects	£2.95
222	Solid State Short Wave Receivers for Beginners	£1.95
223	50 Projects Using IC CA3130	£1.25
224	50 CMOS IC Projects	£2.95
225	A Practical Introduction to Digital ICs	£1.75
226	How to Build Advanced Short Wave Receivers	£2.95
227	Beginners Guide to Building Electronic Projects	£1.95
228	Essential Theory for the Electronics Hobbyist	£2.50
BP1+14	First & Second Books of Transistor Equivalents & Substitutes	£3.50
BP2	Handbook of Radio, TV, Industrial and Transmitting Tube and Valve Equivalents	£0.60
BP6	Engineer's and Machinist's Reference Tables	£1.25
BP7	Radio and Electronic Colour Codes Data Chart	£0.95
BP27	Chart of Radio, Electronic, Semiconductor and Logic Symbols	£0.95
BP28	Resistor Selection Handbook	£0.60
BP29	Major Solid State Audio Hi-Fi Construction Projects	£0.85
BP33	Electronic Calculator Users Handbook	£1.50
BP34	Practical Repair and Renovation of Colour TVs	£2.95
BP36	50 Circuits Using Germanium Silicon and Zener Diodes	£1.50
BP37	50 Projects Using Relays, SCRs and TRIACs	£1.95
BP39	50 (FET) Field Effect Transistor Projects	£1.75
BP42	50 Simple LED Circuits	£1.95
BP44	IC 555 Projects	£2.50
BP45	Projects in Opto-electronics	£1.95
BP48	Electronic Projects for Beginners	£1.95
BP49	Popular Electronic Projects	£2.50
BP53	Practical Electronics Calculations and Formulae	£2.95
BP54	Your Electronic Calculator and Your Money	£1.35
BP56	Electronic Security Devices	£2.50
BP58	50 Circuits Using 7400 Series ICs	£2.50
BP59	Second Book of CMOS IC Projects	£1.95
BP60	Practical Construction of Pre-amps, Tone Controls, Filters and Attenuators	£1.95
BP61	Beginners Guide to Digital Techniques	£1.95
BP62	The Simple Electronic Circuit & Components (Elements of Electronics – Book 1)	£3.50
BP63	Alternating Current Theory (Elements of Electronics – Book 2)	£3.50
BP64	Semiconductor Technology (Elements of Electronics – Book 3)	£3.50
BP65	Single IC Projects	£1.50
BP66	Beginners Guide to Microprocessors and Computing	£1.95
BP67	Counter Driver and Numeral Display Projects	£2.95
BP68	Choosing and Using Your Hi-Fi	£1.65
BP69	Electronic Games	£1.75
BP70	Transistor Radio Fault-finding Chart	£0.95
BP71	Electronic Household Projects	£1.75
BP72	A Microprocessor Primer	£1.75
BP73	Remote Control Projects	£2.50
BP74	Electronic Music Projects	£2.50
BP75	Electronic Test Equipment Construction	£1.75
BP76	Power Supply Projects	£2.50
BP77	Microprocessing Systems and Circuits (Elements of Electronics – Book 4)	£2.95
BP78	Practical Computer Experiments	£1.75
BP79	Radio Control for Beginners	£1.75
BP80	Popular Electronic Circuits – Book 1	£2.95
BP82	Electronic Projects Using Solar Cells	£1.95
BP83	VMOS Projects	£1.95
BP84	Digital IC Projects	£1.95
BP85	International Transistor Equivalents Guide	£3.50
BP86	An Introduction to BASIC Programming Techniques	£1.95
BP87	50 Simple LED Circuits – Book 2	£1.35
BP88	How to Use Op-Amps	£2.95
BP89	Communication (Elements of Electronics – Book 5)	£2.95
BP90	Audio Projects	£1.95
BP91	An Introduction to Radio DXing	£1.95
BP92	Electronics Simplified – Crystal Set Construction	£1.75
BP93	Electronic Timer Projects	£1.95
BP94	Electronic Projects for Cars and Boats	£1.95
BP95	Model Railway Projects	£1.95
BP97	IC Projects for Beginners	£1.95
BP98	Popular Electronic Circuits – Book 2	£2.25
BP99	Mini-matrix Board Projects	£1.95
BP101	How to Identify Unmarked ICs	£0.95
BP103	Multi-circuit Board Projects	£1.95
BP104	Electronic Science Projects	£2.25

BP105	Aerial Projects	£1.95
BP106	Modern Op-amp Projects	£1.95
BP107	30 Solderless Breadboard Projects – Book 1	£2.25
BP108	International Diode Equivalents Guide	£2.25
BP109	The Art of Programming the 1K ZX81	£1.95
BP110	How to Get Your Electronic Projects Working	£1.95
BP111	Audio (Elements of Electronics – Book 6)	£3.50
BP112	A Z-80 Workshop Manual	£3.50
BP113	30 Solderless Breadboard Projects – Book 2	£2.25
BP114	The Art of Programming the 16K ZX81	£2.50
BP115	The Pre-computer Book	£1.95
BP117	Practical Electronic Building Blocks – Book 1	£1.95
BP118	Practical Electronic Building Blocks – Book 2	£1.95
BP119	The Art of Programming the ZX Spectrum	£2.50
BP120	Audio Amplifier Fault-finding Chart	£0.95
BP121	How to Design and Make Your Own P.C.B.s	£1.95
BP122	Audio Amplifier Construction	£2.25
BP123	A Practical Introduction to Microprocessors	£1.95
BP124	Easy Add-on Projects for Spectrum, ZX81 & Ace	£2.75
BP125	25 Simple Amateur Band Aerials	£1.95
BP126	BASIC & PASCAL in Parallel	£1.50
BP127	How to Design Electronic Projects	£2.25
BP128	20 Programs for the ZX Spectrum and 16K ZX81	£1.95
BP129	An Introduction to Programming the ORIC-1	£1.95
BP130	Micro Interfacing Circuits – Book 1	£2.25
BP131	Micro Interfacing Circuits – Book 2	£2.25
BP132	25 Simple Shortwave Broadcast Band Aerials	£1.95
BP133	An Introduction to Programming the Dragon 32	£1.95
BP134	Easy Add-on Projects for Commodore 64, Vic-20, BBC Micro and Acorn Electron	£2.95
BP135	Secrets of the Commodore 64	£1.95
BP136	25 Simple Indoor and Window Aerials	£1.75
BP137	BASIC & FORTRAN in Parallel	£1.95
BP138	BASIC & FORTH in Parallel	£1.95
BP139	An Introduction to Programming the BBC Model B Micro	£1.95
BP140	Digital IC Equivalents and Pin Connections	£5.95
BP141	Linear IC Equivalents and Pin Connections	£5.95
BP142	An Introduction to Programming the Acorn Electron	£1.95
BP143	An Introduction to Programming the Atari 600/800 XL	£1.95
BP144	Further Practical Electronics Calculations and Formulae	£4.95
BP145	25 Simple Tropical and MW Band Aerials	£1.75
BP146	The Pre-BASIC Book	£2.95
BP147	An Introduction to 6502 Machine Code	£2.50
BP148	Computer Terminology Explained	£1.95
BP149	A Concise Introduction to the Language of BBC BASIC	£1.95
BP150	An Introduction to Programming the Sinclair QL	£1.95
BP152	An Introduction to Z80 Machine Code	£2.75
BP153	An Introduction to Programming the Amstrad CPC 464 and 664	£2.50
BP154	An Introduction to MSX BASIC	£2.50
BP155	International Radio Stations Guide	£2.95
BP156	An Introduction to QL Machine Code	£2.50
BP157	How to Write ZX Spectrum and Spectrum + Games Programs	£2.50
BP158	An Introduction to Programming the Commodore 16 and Plus 4	£2.50
BP159	How to Write Amstrad CPC 464 Games Programs	£2.50
BP161	Into the QL Archive	£2.50
BP162	Counting on QL Abacus	£2.50
BP163	Writing with QL Quill	£2.50
BP164	Drawing on QL Easel	£2.50
BP169	How to Get Your Computer Programs Running	£2.50
BP170	An Introduction to Computer Peripherals	£2.50
BP171	Easy Add-on Projects for Amstrad CPC 464, 664, 6128 and MSX Computers	£2.95
BP173	Computer Music Projects	£2.95
BP174	More Advanced Electronic Music Projects	£2.95
BP175	How to Write Word Game Programs for the Amstrad CPC 464, 664 and 6128	£2.95
BP176	A TV-DXers Handbook	£5.95
BP177	An Introduction to Computer Communications	£2.95
BP178	An Introduction to Computers in Radio	£2.95
BP179	Electronic Circuits for the Computer Control of Robots	£2.95
BP180	Computer Projects for Model Railways	£2.95
BP181	Getting the Most from Your Printer	£2.95
BP182	MIDI Projects	£2.95
BP183	An Introduction to CP/M	£2.95
BP184	An Introduction to 68000 Assembly Language	£2.95
BP185	Electronic Synthesiser Construction	£2.95
BP186	Walkie-Talkie Projects	£2.95
BP187	A Practical Reference Guide to Word Processing on the Amstrad PCW 8256 and PCW 8512	£5.95
BP188	Getting Started with BASIC and LOGO on the Amstrad PCW 8256 and PCW 8512	£6.95
BP189	Using Your Amstrad CPC Disc Drives	£2.95
BP190	More Advanced Electronic Security Projects	£2.95
BP191	Simple Applications of the Amstrad CPCs for Writers	£2.95
BP192	More Advanced Power Supply Projects	£2.95
BP193	Starting LOGO	£2.95
BP194	Modern Opto Device Projects	£2.95
BP195	An Introduction to Communications and Direct Broadcast Satellites	£3.95
BP196	BASIC & LOGO in Parallel	£2.95